Mirella Gamboa Narváez
José Luis Rios Flores
Becky Elizabeth Ríos Arredondo

# Efficiency and productivity of water

**Mirella Gamboa Narváez**
**José Luis Rios Flores**
**Becky Elizabeth Ríos Arredondo**


# Efficiency and productivity of water

## Used in the production of pecan nut (Carya illinoensis) versus apple (Malus domestica) in Chihuahua, Mexico


**ScienciaScripts**

**Imprint**

Any brand names and product names mentioned in this book are subject to trademark, brand or patent protection and are trademarks or registered trademarks of their respective holders. The use of brand names, product names, common names, trade names, product descriptions etc. even without a particular marking in this work is in no way to be construed to mean that such names may be regarded as unrestricted in respect of trademark and brand protection legislation and could thus be used by anyone.

Cover image: www.ingimage.com

This book is a translation from the original published under ISBN 978-620-3-88753-2.

Publisher:
Sciencia Scripts
is a trademark of
Dodo Books Indian Ocean Ltd. and OmniScriptum S.R.L publishing group

120 High Road, East Finchley, London, N2 9ED, United Kingdom
Str. Armeneasca 28/1, office 1, Chisinau MD-2012, Republic of Moldova, Europe
Printed at: see last page
**ISBN: 978-620-7-67535-7**

# Contents

SUMMARY

The objectives were to determine the physical (PFA), economic (PEA) and social (PSA) productivity, as well as the physical (EFA), economic (EEA) and social (ESA) efficiency of water used in the production of pecan nut (*Carya illinoensis*) and contrast it against apple (*Malus domestica*) in the state of Chihuahua in 2019, both crops produced under medium technology use (MT) conditions. Models of productivity and water efficiency in the agricultural sector were used, feeding these models with production data at commercial level from SIAP. The results show that the PFA, PEA and PSA were (always MT walnut vs MT apple): 0.15 and 2.28 kg m$^{-3}$ , USD 126,897 and USD 619,570 profit per hm$^3$ and 9 and 23 jobs per hm$^3$ , while the EFA, EEA and ESA were: 6,603 and 438 litres kg$^{-1}$ , 7.88 and 1.61 m$^3$ of water used per USD of profit and 113,143 and 42,562 m$^3$ of water used per job. The indicators show that in physical, economic and social terms, the use of water in production is less productive (therefore less efficient) in walnut, because when using the same volume of water used by apple, it produces only 6.63% of the physical product, 20.48% of the profit and 37.6% of the employment generated by apple. The average EAP of the MT apple (USD 619,570 hm$^{-3}$ ) for Chihuahua was only surpassed by 72% by the MT olive tree (USD 619,570 hm ) for Chihuahua.

Spain; BT, MT and AT apples (produced with low "B", medium "M" and high "A" technology) from Cuauhtemoc, Chihuahua, as well as BT and AT apples from Canatlan and Santiago Papasquiaro, Durango, had lower EAP indices, but not PES, as the MT apple and MT walnut analysed (9 and 23 jobs hm$^{-3}$ respectively), were outperformed by the social productivity of water in various crops such as the BT apples of Cuauhtemoc, Chihuahua and Canatlan and Santiago Papasquiaro, Durango.

**Keywords:** efficiency, virtual water, sustainability, water footprint.

# I. INTRODUCTION

The apple (*Malus domestica* L.) is one of the most consumed foods by the human being, likewise, it is estimated that it is one of the foods that occupy a greater percentage in the daily expenditure of the people when providing vegetables in their food, on the other hand, although the pecan nut (*Carya illinoensis*) is not a food as usual as the apple, in the last years its production has increased notably, both crops have increased sensibly in their demand *per capita*, in the case of the apple is located at the moment in 8.8 kg per person per year[1] , while in the case of walnuts the per capita consumption is 0.5 kg per person per year in Mexico[2] , and both crops demand relatively much water in their production depending on where they are produced and in which production system they are produced, on the other hand, in 1970 Mexico had a population of 51,493,565 people, and by 2018 in the country there were already 124,738,00 inhabitants[3] [4] , so, if we consider a water footprint of 822 litres of water per kg of apple[5] and of 16.19 $m^3$ $kg^{-1}$ in the case of pecan nuts[6] , then, of the annual water footprint of each Mexican, that is, of the total direct and virtual water consumed by each Mexican, by the simple annual

---

[1] SIAP, 2017. Apple: Mexico produced 716,930 tonnes of apples in 2016. Available at:
https://www.gob.mx/siap/articulos/manzana-mexico-produjo-716-930-toneladas-en-2016?idiom=es
[2] **Ortiz, Ramos Daniel, 2016**. The production and foreign trade of the walnut (*Carya ilinoinensis*) in Mexico. Professional thesis of Licenciado en Comercio Internacional. Universidad Autonoma Agraria Antonio Narro, Unidad Saltillo. Saltillo, Coahuila Mexico Pag. 33. Available at:
http://repositorio.uaaan.mx:8080/xmlui/bitstream/handle/123456789/8211/T19328%20ORTIZ%20R AMOS,%20DANIEL.pdf?sequence=1
[3] Mexico-Population, 2018. Expansion/ datos macro.com. Available at:
https://datosmacro.expansion.com/demografia/poblacion/mexico
[4] **Evolution of the world population**. The market economy: virtues and innovations. Demograffa. 2019. Available at: http://www.juntadeandalucia.es/averroes/centros-tic/14002996/helvia/aula/archivos/repositorio/250/271/html/economia/2/evolucion.htm. Accessed September 10, 2019.
[5] **HOEKSTRA, A. Y. and HUNG, P.Q. (2005)**. "Globalization of water resources: international virtual water flows in relation to crop trade". Global Environmental Change, 15, pp. 45-56.
[6] **Cifuentes Rodriguez, Reynau, 2017**. Agricultural productivity of water, soil, capital and labour in the cultivation of walnut (*Carya illinoensis*) in San Pedro, Coahuila. Professional thesis. Department of Irrigation. Universidad Autonoma Agraria Antonio Narro, Laguna Unit. Torreon, Coahuila, Mexico.

consumption of apples and nuts, each Mexican is spending 15,328.6 litres of water (7,233.6 litres of water for apple consumption and 8,095 litres of water per person per year for pecan nut consumption), which, with the population in 2018, implies that 1,912.058907 million $m^3$ per year were invested just to supply the population with the apple and pecan nut that they consume annually.

Thus, the above is the first of the faces or facets of the problem to be addressed within which this work is inserted: the growth in demand for food of agricultural origin, as well as demographic growth, which exert enormous pressure on the use of water, an extremely scarce resource, since of the total amount of water on the planet, only 0.26% of the 1,386 million $km^3$ of water on the planet is ***available fresh water***, the rest is made up of 2.24% of ***unavailable fresh water***, as it is found in the polar ice caps, permafrost, glaciers and extremely deep waters, at depths of more than one km, inaccessible in economic terms with current technology, and 97.5% is ***salt water*** from the seas and oceans[7].

The second aspect or facet of the problem within which this work is located, is that the agricultural sector is the main user of available fresh water on the planet, on average worldwide, agriculture and livestock consume 70% of the available fresh water, 20% is for industrial use and 10% for domestic consumption, although as all averages vary from country to country, for example, in India, the main consumer of fresh water (with 761 thousand $hm^3$ /year) 90.4% is used for agriculture, 2.2% for industry and 7.China is the second largest consumer of freshwater, with 607.8 thousand $hm^3$ /year, of which 64.5% is for agricultural use, the USA is the third largest consumer of freshwater, with 485.6 thousand $hm^3$ /year, of which only 36.1 is used by agriculture, 51.Mexico is the world's 7th largest

---

[7] Aqua Foundation. Quantity of drinking water, source of life. Available from:
https://www.fundacionaquae.org/wiki-aquae/datos-del-agua/cantidad-de-agua-potable-fuente-de- vida/
Accessed 10 September 2019.

consumer of freshwater, with 85.66 thousand hm$^3$ /year, of which 76.3% (i.e. above the world average) is used by agriculture, 9.1% by the industrial sector and 14.6% is for public supply[8] .

As the state of Chihuahua is the main national producer of both apple and pecan nuts, it is important to study how the scarce water resource is being used in their production. Based on the above, it should be clear that what is analysed in this paper is neither the apple nor the walnut, but simply the physical, economic and social productivity of water use in the production of pecan and apple in the state of Chihuahua.

In the case of Mexico in particular, with 0.1% of the world's total available freshwater, agriculture and livestock consume 76.3% of the available fresh water (remember that the world average is 70%), but in addition, the third facet of the problem in which this work is inserted, is that being little fresh water available in the world, and being agriculture and livestock the one who consumes more, also *uses it inefficiently*, 57% of the water consumed is lost due to various causes such as evaporation, but it is also due to inadequate, inefficient, obsolete and neglected irrigation infrastructure. Worse still, irrigated agriculture generates only 42% of national production[9] .

The above suggests that numerical indicators are needed to indicate how **productive** (i.e. how much *physical or economic or social output* is achieved per unit volume of water) and **efficient** (i.e. how much water is used per unit physical or economic or social output) the water use of pecan nut (Carya illinoensis) crops is compared to apple (Malus domestica) crops grown under Median conditions, when both crops are grown under Median conditions, and how much water is used per unit

---

[8] FAO (Food and Agriculture Organisation of the United Nations), Agriculture and Consumer Protection Department, 2005. Water use in agriculture. Available at:
http://www.fao.org/ag/esp/revista/0511sp2.htm

[9] Agua.org.mx Fondo para la Comunicacion y la Educacion Ambiental A.C. Overview of water in Mexico. Available at: https://agua.org.mx/cuanta-agua-tiene-mexico/. Last accessed 21 January, 2020.

volume of water, (i.e. how much water is used per unit of physical, economic or social output) is the water use of pecan nut (*Carya illinoensis*) cultivation compared to apple (*Malus domestica*) cultivation, both crops grown under Medium Technology Use (hereafter MT) conditions, so that based on these index numbers, those who make macroeconomic decisions on water use and allocation among different alternatives to which water can be subjected, can *optimise* their use of water.

# II. OBJECTIVES AND HYPOTHESES
## II.1. General and specific objectives

The general objectives of this study were to determine the productivity and efficiency of water used in the production of pecan nut (*Carya illinoensis*) produced under conditions of medium use of "MT" technology and to contrast it against the corresponding indicators of productivity and efficiency of water used in the production of MT apple (*Malus domestica*) at the level of the state of Chihuahua, Mexico.

## II.2. Hypothesis
### II.2.1.     First hypothesis

At an aggregate level of the whole state of Chihuahua, MT pecan nut (*Carya* illinoensis) in relation to apple (*Malus domestica*), will have a <u>lower ***physical***</u> water productivity "PFA" (measured PFA in kg of product per $m^3$ of water used in production, abbreviated as kg $m^{-3}$ ), i.e. the pecan nut produces less biomass per $m^3$ of water used in production than the MT apple crop (*Malus domestica*).

### II.2.2.     Second hypothesis

At an aggregate level of the whole state of Chihuahua, MT pecan nut (*Carya illinoensis*) in relation to apple (*Malus domestica*), will have a <u>lower ***economic***</u> water productivity "EAP" (measured EAP in USD profit per m3 of water used in production, abbreviated as USD $m^{-3}$ ), i.e. the pecan nut produces less USD profit per $m^3$ of water used in production than the MT apple crop (*Malus domestica*).

### II.2.3.     Third hypothesis

At an aggregate level of the whole state of Chihuahua, MT pecan nut (*Carya illinoensis*) in relation to apple (*Malus domestica*), will have a <u>lower ***social***</u> water productivity "PES" (measured in jobs generated per $hm^3$ of water used in production, abbreviated as jobs $hm^{-3}$ ), i.e. pecan nut produces fewer jobs per $hm^3$ of water used in production than MT apple (*Malus domestica*).

# III. LITERATURE REVIEW
## III.1. Characteristics of apple (*Malus domestica*) and pecan nuts (Carya illinoensis)
### pecan nut (*Carya illinoensis*)

In general, the colour shades assumed by apples in the world range from 530 to 740 nanometres, according to the spectrum visible to the human eye, see figure 1.

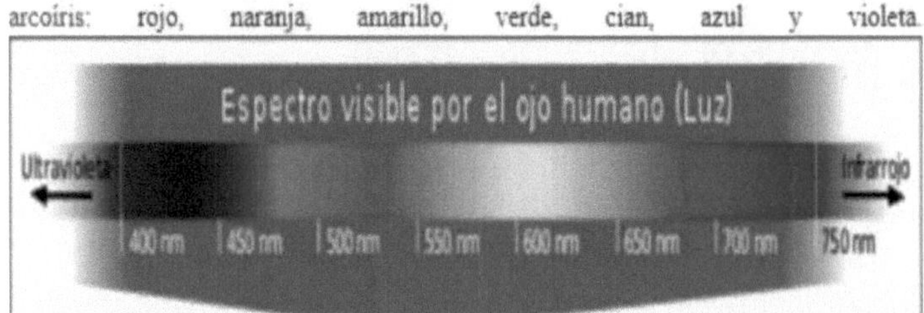


**Figure 1. Spectrum visible to the human eye and colour range of the apple.**

The apple is a fruit of firm structure, it comes from the receptacle of a flower, its colour can vary: intense red, pale red, orange, intense green, pale green and yellow, depending on the variety.

The Gala variety is bicoloured, crunchy and with sweet aromas, it is one of the sweetest apples, it has orange and pink stripes on a yellow background, it is one of the most consumed varieties in the world. It is widely used in salads and jams.

**Figure 2. Different apple colourings.**

Red Delicious or Red Apple variety: crunchy and slightly sweet, excellent for use in salads and delicious smoothies, the heart-shaped fruit is bright red in colour. It may help to protect the urinary tract, cleanse the bowel and contains antioxidants. The common season to enjoy this apple is all year round.

Golden Delicious or Yellow Apple: it is the ideal choice for any recipe, it is sweet and soft with a tender skin, and its pulp remains white for longer than other apples after cutting them. It is recommended for salads, soups, drinks, cakes or any elaboration for refrigeration. It is one of the first varieties to be selected in the United States.

Granny Smith or Green Apple: Known for their delicious tart taste and crunchiness, we are surprised by the popularity of this apple. Granny Smith apples are great for any kind of recipe, such as salads, mashed apples, baked goods, refrigerated dishes and more. Sliced, with a little spice, lemon and salt they are the perfect snack for watching football games, as well as making the perfect centrepieces for a wedding or ceremony.

In addition to the specific properties of each colour, it has been discovered that apples can bring health to the intestines, as well as stimulate the immune system by increasing the amount of good intestinal bacteria, for this and more they are considered a super fruit.

Apples, in general, regardless of variety and colour, have anti-diarrhoeal, laxative, diuretic, depurative, lipid-lowering, nervous system toning and hypotensive properties. Apples are known for their high water content, around 85%, which makes them very refreshing and hydrating. Moreover, the sugars present in this fruit are mainly fructose (fruit sugar), and to a lesser extent glucose and sucrose, which are sugars that are quickly assimilated by the organism[11] .

Apples are rich in antioxidants, above all they have a large amount of Vitamin E and Vitamin C; rich in fibre, which

---

[11] Ecosarga: The properties of apples according to their colour. Available at:
https://www.frutadelasarga.com/blog/las-propiedades-de-las-manzanas-segun-su-color

improves intestinal transit and, in their mineral content, it is worth highlighting their high potassium and iron content, which makes them a fruit suitable for all types of people.

At present, there are about 4000 known plant pigments in our food. Each colour group represents different phytochemicals. Phytochemicals are substances that are linked to a lower incidence of cancer, heart disease, among others. Phytochemicals are chemical elements found in foods of plant origin, but they are neither nutrients nor macronutrients, nor are they included in the group of vitamins or minerals. Therefore, they have no energetic or nutritional function, but nevertheless provide various beneficial functions. This is why foods containing phytochemical substances are called functional foods, as in addition to the nutritional component, they also provide other health benefits, such as antioxidant and anti-cancer properties. In this sense, phenols and terpenes stand out, including flavonoids and carotenoids, which help to reduce inflammation and act as protectors in cardiovascular diseases such as heart attacks, angina pectoris, arterial hypertension, arteriosclerosis, etc. Some also help to maintain the immune system[12] .

### III.1.1.    The colours of apples and their properties

**Red apples.** They belong to the group of red phytonutrients (lycopene, anthocyanins and ellagic acid) which provide optimal nutrition and well-being for the heart, cells and skin, improve immunity and help to counteract the effects of stress and fatigue. In addition to red apples, these compounds are found in grapefruit, cherries, red grapes, red pears, pomegranates, raspberries, watermelon, red peppers, radishes, purple onions and tomatoes.

**Yellow apples.** *They are associated with protection against some types of cancer,* support eye health (particularly night

---

[12] What are phytochemicals?          Available at: https://www.espn.com.mx/espn-run/note/_/id/2722475/nutrition-what-is-phytochemicals

vision), cardiovascular and immune system protection. Normally orange or yellow foods contain Vitamin C and the phytochemicals carotenoids, lutein and terpenes. In oranges, melons, tangerines, zucchini, pineapple, papaya, corn, lemon, carrots, pumpkin, etc. we find these nutrients.

**Green apples.** They are also related to the protection of some types of cancers, maintenance of bones and teeth and also provide benefits for the eyesight. Polyphenols and isoflavones are the best known phytochemicals. *The darker the green colour, the higher the content of protective substances.* Kiwifruits, green pepper, parsley, broccoli, spinach, asparagus, cabbage and its derivatives, cabbage, cucumbers, etc. are the most important foods in terms of vitamin K1 content.

In relation to nuts, they are among the most complete foods in the world. Among the many properties of walnuts there are 6 very remarkable[13] :

**1.** They reduce the level of bad LDL cholesterol and hypertension.

**2.** Prevents the risk of cardiovascular diseases such as heart attack and angina pectoris.

**3.** Prevents the development of arteriosclerosis.

**4.** They are great food for our brains.

**5.** Because of their antioxidant power and vitamins, they are beneficial for our skin.

**6.** Due to its satiating capacity, it is recommended for preventing or treating obesity.

### III.1.2.　Benefits and properties of walnuts

The main properties of walnuts are that they are rich in Omega 3 fats which help to lower our cholesterol and prevent bad circulation.

Regular consumption of walnuts, about 5 walnuts 5 times a week, can reduce the risk of cardiovascular diseases such as

---

[13] The six surprising properties of walnuts. Available at: https://comefruta.es/las-6- surprising-properties-of-nuts

heart attack or angina pectoris by 50%.

According to some studies, walnuts can reduce the level of LDL (bad cholesterol) in the blood to a greater extent than olive oil, making walnuts one of the best foods for fighting cholesterol.

Other significant properties of walnuts are:

. Lowers hypertension (garlic is another natural remedy for hypertension).

. Prevents the onset of arteriosclerosis.

. Suitable for diabetics due to their low carbohydrate content.

### III.2. Apple and pecan nut production in Chihuahua

In 2018 the world apple production amounted to 74,350,175 tons, China contributed 56%, the European Union contributed 14% and the U.S.A. 7%, the rest of the world contributed the remaining 23%, in that year, 2018, according to table 1, Mexico contributed 659,451 tons, equivalent to 0.89% of world production.

## Table 1. Statistics of apple production in Mexico 2009-2018

| Ano | Harvested area (ha) | Production (ton) | Value (MX$ millions nominal) | Value (MX$ millions constant December 2018) | PMR/ton (MX$ nominal) | PMR/ton (MX$ constant 2018) | Yield (ton/ha) |
|---|---|---|---|---|---|---|---|
| 2009 | 56,992 | 561,493 | $2,333 | $3,835 | $4,155 | $6,830 | 9.852 |
| 2010 | 57,743 | 584,655 | $3,253 | $4,817 | $5,564 | $8,239 | 10.125 |
| 2011 | 56,845 | 630,533 | $3,123 | $4,192 | $4,953 | $6,648 | 11.092 |
| 2012 | 58,451 | 375,045 | $3,009 | $3,870 | $8,023 | $10,318 | 6.416 |
| 2013 | 59,199 | 858,608 | $4,265 | $5,447 | $4,967 | $6,344 | 14.504 |
| 2014 | 55,447 | 716,865 | $4,206 | $5,068 | $5,867 | $7,069 | 12.929 |
| 2015 | 55,121 | 750,325 | $4,322 | $5,119 | $5,760 | $6,822 | 13.612 |
| 2016 | 54,248 | 716,930 | $4,659 | $5,038 | $6,499 | $7,027 | 13.216 |
| 2017 | 53,619 | 714,149 | $6,231 | $6,395 | $8,725 | $8,955 | 13.319 |
| 2018 | 49,493 | 659,451 | $7,780 | $7,780 | $11,798 | $11,798 | 13.324 |
| TAC (2018/2009) | -1.4 | 1.6 | | 7.3 | | 5.6 | 3.1 |

**Source:** Own elaboration, based in figures from: 2018. Statistics of the apple in Mexico. https://blogagricultura.com/estadisticas-manzana-mexico/. Production value in constant MX$ of December 2018 is own elaboration.

Table 1 shows that between the year 2009 and the year 2018, production grew, in absolute terms, 97,958 tons (17.4%) even though apple area decreased by 7,499 ha (13.2%) in the period, in terms of annual growth rate (AAGR), Table 1 shows that while area decreased at a rate of 1.4% each year, production was increasing at a rate of 1.6% each year, and furthermore, the value of production, already expressed in constant MX$ as of December 2018, was growing at a rate of 7.3% per annum, the answer is in the last two columns: the price of apples and the yield per hectare, as for example, the yield per hectare, going from 9.852 to 13.324 ton ha$^{-1}$ grew at a rate of 3.1% each year, and in conjunction with the real prices of apples increasing at a rate of 5.6% annually, necessarily resulted in the value of production, through a combined effect of prices and yields, growing at 7.3% per annum, from 3,835 to 7,780 million constant pesos.

Table 2 shows the top ten apple producing states in Mexico. It is observed that in 2018 the state of Chihuahua, with 25,245 ha harvested in 2018, contributed 52.2% of the total harvested area, with 569,580 tonnes it contributed 87.4% of the total and represented almost 94 cents of every peso of value generated by the apple, and with respect to the national total, the apple of Chihuahua represented 51% of the national harvested area of apple, 86.4% of the national production and generated 93 cents of every peso of value produced nationally by the crop.

## Table 2. Apple-producing states in Mexico in 2018

| State | Harvested area (ha) | Production (ton) | Value (MX$ millions) |
|---|---|---|---|
| 1 Chihuahua | 25,245 | 569,580 | 7,235 |
| 2 Puebla | 7,814 | 35,713 | 162 |
| 3 Durango | 6,749 | 11,146 | 75 |
| 4 Coahuila | 3,263 | 10,165 | 74 |
| 5 Veracruz | 1,329 | 9,236 | 59 |
| 6 Zacatecas | 1,014 | 4,442 | 26 |
| 7 Chiapas | 814 | 3,370 | 26 |
| 8 Hidalgo | 804 | 3,353 | 22 |

| | | | |
|---|---|---|---|
| **9 Nuevo Leon** | 706 | 2,760 | 21 |
| **10 Oaxaca** | 597 | 2,281 | 21 |
| **Total** | 48,335 | 652,046 | 7,721 |
| **Chihuahua/total 10 states** | 52.2% | 87.4% | 93.7% |
| **Chihuahua/national total** | 51.0% | 86.4% | 93.0% |

**Source:** Sta^sticas de la manzana en Mexico. https://blogagricultura.com/estadisticas-manzana-mexico/. The percentages of Chihuahua's participation are the author's own elaboration based on table 1 and 2.

The remaining nine apple-producing states, while accounting for 49% of the harvested area, only account for 13.7% of the annual physical production and 7% of the national value of apple production, according to table 2.

Figure 3 shows the evolution of the harvested area and the annual physical production of the pecan nut (*Carya illinoensis*) crop in Mexico between the year 2000 and the year 2017, observing that at the beginning of the indicated period of time, the harvested area was only about 50 thousand hectares and by 2017 it had become 123,266 ha, while the annual physical production, going from 60 to 150.349 thousand tonnes per year, which suggests that the annual growth rates of both harvested area and production advanced at a rate of 5.5% each year, which in absolute terms, is equivalent to increasing each year a little more than 5 thousand tonnes of walnuts, which is not a negligible figure. The annual growth rate of the physical production of walnuts in Mexico, 5.5%, is a little more than three times the annual growth rate of apple production, 1.6%, shown in table 1.

Figure 3. Harvested area (ha) and production of pecan nuts in Mexico, 2000-2017[13] .[14]

Figures 3, 4 and 5 show, respectively, the evolution of the annual physical production of pecan nuts in Mexico, the U.S.A. and at the aggregate level for both countries in the period 2000 to 2017.

---


[1413] *Comite Mexicano del Sistema Productivo Nuez A.C. 2018*. Y Alderete y Socios, Industrial Consultancy. 2018. Pecan nut strategic study. Actualization 2018. Available *at:* http://comenuez.com/wp-content/uploads/2018/assets/estudio-estrategico-nuez-pecanera-- 2018.pdf. Last accessed on February 12, 2020.

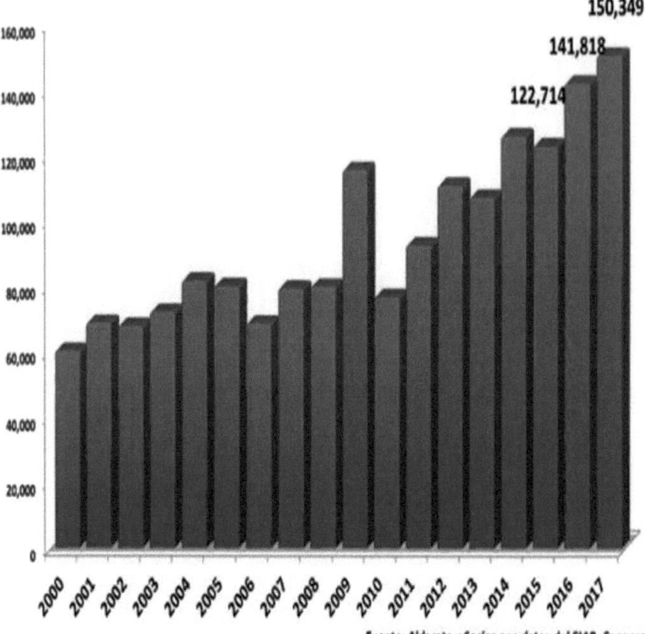


*Source: Alderete y Partners with data from 5IAP- Sagarpa*

**Figure 4. Pecan nut (*Carya illinoensis*) production (ton) in Mexico 2000 2017[15]**

From 2015 Mexico, with 122,714 tons per year (see Figure 4), Mexico became the world's leading producer of walnuts, surpassing the U.S. with 115,448 tons (see Figure 5), and by 2017, Mexico produced 150,349 tons, while the U.S. produced 125,829 tons (see Figure 5).U.A. produced 125,829 tons (see figure 5), year in which with a joint production of just over 276 thousand tons in both countries, Mexico contributed 54% of the joint production and the U.S. contributed 46% (see figure 5).

**Pecan Nut Production USA 2000-2017**

---

[15] ***Comite Mexicano del Sistema Productive Nuez A.C. 2018. Y Alderete y Socios, Industrial Consultancy. 2018.*** *Pecan nut strategic study. Actualization 2018. Available **at:*** http://comenuez.com/wp-content/uploads/2018/assets/estudio-estrategico-nuez- pecanera--2018.pdf. *Last accessed on February 12, 2020.*

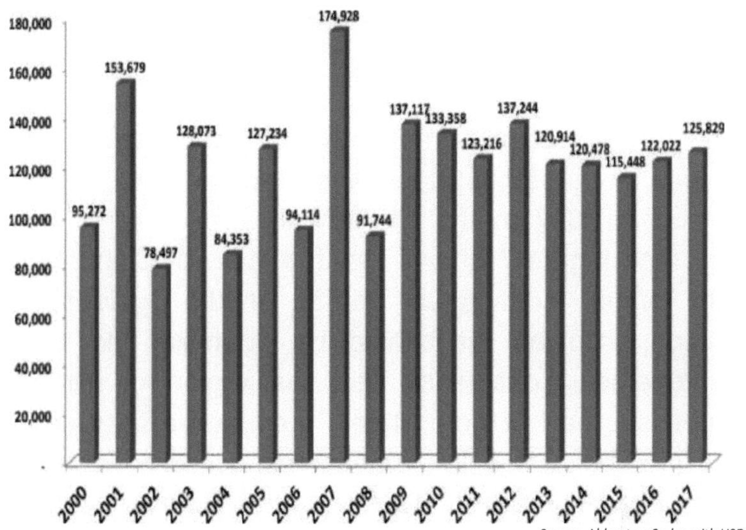


**Figure 5. Pecan nut (*Carya illinoensis)* production in the USA, 2000-2017[15] .**

[15] *Comite Mexicano del Sistema Productive Nuez A.C. 2018. Y Alderete y Socios, Industrial Consultancy. 2018. Pecan nut strategic study. Actualization 2018.*

Figures 4 and 5 show that while in Mexico walnut production fell in 2007 and rose sharply in 2010, in the U.S. 2007 was an exceptionally high production year, with a production of 174,928 tons, but not in 2010, when only 133,358 tons of walnuts were harvested.

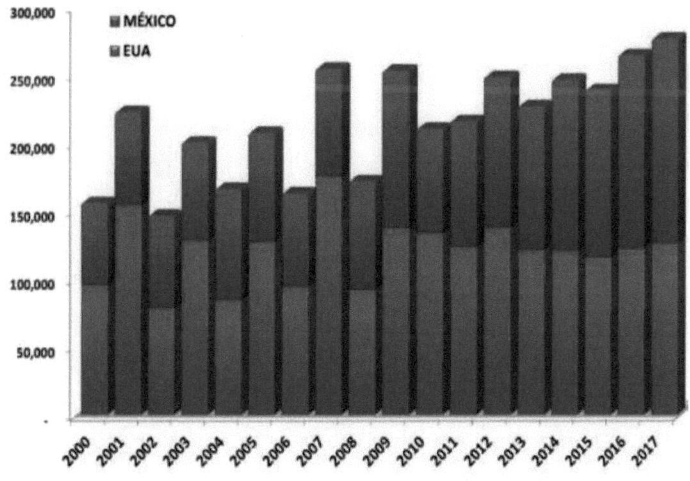


**Figure 6. Joint pecan nut production between Mexico and the USA: of the 276**

**thousand tonnes in 2017, 54% was Mexican pecans and 46% was from the USA. Available at: http://comenuez.com/wp-content/uploads/2018/assets/estudio-pecan-nut-strategy--2018.pdf[16]**

Figure 7 shows the breakdown of the annual physical production of the 150,349 tonnes (shown in Figure 4) produced in Mexico in 2017 in the top ten pecan-producing states in Mexico. Ash Figure 7 shows that the state of Chihuahua with 96,926 tonnes contributed 64.5% of the national pecan production, followed far behind by Coahuila and Durango, which together, considering that the region of La Laguna encompasses part of both states, produced a total of 24,443 tonnes (15,962 and 8,481 tonnes respectively) equivalent to 16.3%, while Sonora, with 19,719 tonnes contributed 13.1% of the national production, ash these four states produced 93.8% (141,088 tons), while the remaining six of the top ten walnut producing states in Mexico, being those six states: Nuevo Leon, Hidalgo, San Luis Potosi The state of Mexico, Aguascalientes and Oaxaca, only contributed with 8,274 tons, 5.5%, so the remaining 0.7% of the national production was contributed by the remaining 22 states of the Mexican Republic.

---

[16] *Available **at:** http://comenuez.com/wp-content/uploads/2018/assets/estudio-estrategico-nuez-pecanera--2018.pdf. Last accessed on 12 February, 2020.*

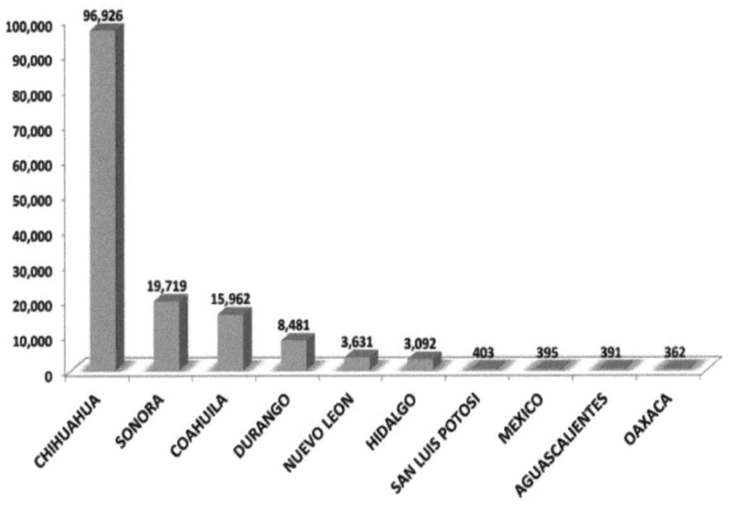



**Figure 7. Pecan nut production (ton) in the main Mexican producing states in 2017. Available at: http://comenuez.com/wp-content/uploads/2018/assets/estudio-estrategico-nuez-pecanera--2018.pdf**

The apparent consumption of walnuts in Mexico (the sum of production plus imports minus exports), as an indicator of the evolution in the demand for walnuts, as shown in Figure 8, has increased from 49,384 tonnes per year in 2009 to 55,026 tonnes per year in 2017, growing at a rate of 1.2% per year, also, Figure 8 shows that the trade balance (exports minus imports) of walnuts in Mexico, in physical terms, has increased from 65,966 tonnes in 2009 to 73,323 tonnes in 2017, increasing each year at a rate of 1.2% per year.

73,323 tonnes in 2017, increasing each year at a rate of 1.2 % per year.

| CONSUMO APARENTE DE MEXICO (TONS) | | | | | | | | | |
|---|---|---|---|---|---|---|---|---|---|
| | 2009 | 2010 | 2011 | 2012 | 2013 | 2014 | 2015 | 2016 | 2017 |
| PRODUCCION | 115,350 | 76,627 | 92,359 | 110,605 | 105,542 | 125,412 | 122,714 | 141,818 | 150,349 |
| EXPORTACIONES | 87,662 | 91,231 | 67,456 | 70,474 | 79,750 | 102,118 | 95,384 | 119,431 | 99,181 |
| CON CASCARA | 41,016 | 19,454 | 24,587 | 25,327 | 23,537 | 32,551 | 34,904 | 43,302 | 23,356 |
| SIN CASCARA | 46,646 | 71,777 | 42,868 | 45,147 | 56,213 | 69,567 | 60,480 | 76,129 | 75,825 |
| IMPORTACIONES | 21,696 | 22,045 | 20,870 | 24,506 | 23,983 | 26,187 | 29,087 | 29,084 | 25,858 |
| CON CASCARA | 18,135 | 15,963 | 16,078 | 17,356 | 18,127 | 20,448 | 23,074 | 25,262 | 22,176 |
| SIN CASCARA | 3,561 | 6,082 | 4,792 | 7,150 | 5,856 | 5,738 | 6,013 | 3,822 | 3,682 |
| SALDO COMERCIAL | 65,966 | 69,186 | 46,586 | 45,968 | 55,767 | 75,932 | 66,297 | 90,346 | 73,323 |
| ALMACENAMIENTO ESTIMADO | | | | | | | | | 22,000 |
| CONSUMO APARENTE | 49,384 | 7,441 | 45,773 | 64,637 | 49,775 | 49,480 | 56,417 | 51,471 | 55,026 |

**Figure 8. Apparent consumption of pecan nuts in Mexico.**
Available at: http://comenuez.com/wp-content/uploads/2018/assets/estudio-estrategico-nuez-pecanera--2018.pdf

The apparent consumption is a clear sign that the pecan nut market in Mexico is expanding, which represents a business opportunity for walnut producers, as the demand for this nut is clearly increasing, as shown in figure 8.

The apparent consumption of the three most important nuts: cashew, pecan and hazelnut in the United States of America, is shown in Figure 9, from which it can be observed that although the apparent consumption has been slightly higher for the cashew nut, its behaviour, trend and values have been very similar for the cashew and pecan nuts, which for 2017 was around 110 thousand tonnes for the pecan nut and just over 160 thousand tonnes for the cashew nut.

**APPARENT CONSUMPTION. The apparent consumption of the pecan nut is very similar to that of the WALNUT. Recently, both intakes have increased. Hazelnut has lower consumption levels.**

20

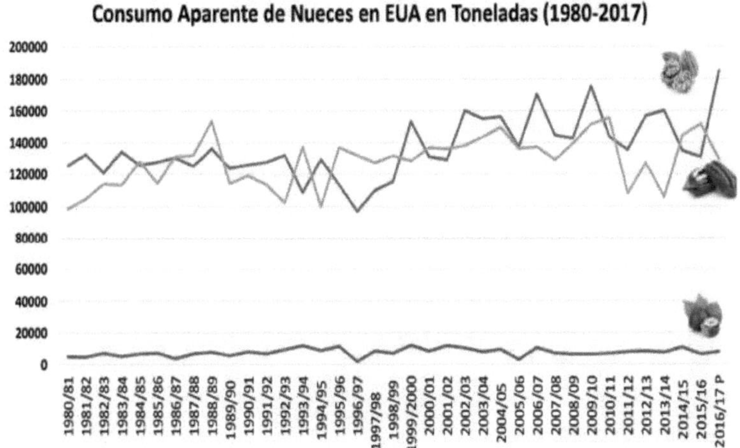


**Figure 9. Apparent consumption of various types of walnuts in the USA in tonnes (1980**
**2017).**
**Available at: http://comenuez.com/wp-content/uploads/2018/assets/estudio-estrategico-nuez-pecanera--2018.pdf**

In terms of trend, Figure 9 shows that in the case of the castile nut, the trend has had a greater positive slope than in the case of the pecan nut, whose slope, also increasing, also positive, has been relatively less marked in relation to the castile nut.

The third nut, the hazelnut, whose apparent consumption growth has been rather stationary, has never after 1980 reached an apparent consumption of around 20,000 tonnes per year, but has been close to 10,000 tonnes of apparent consumption per year (see Figure 9).

# EXPORT POSITIONING 2016

**Figure 10. The pecan nut (*Carya illinoensis*) is the twelfth largest agri-food product generating export value in Mexico.**
**Available at: http://comenuez.com/wp-content/uploads/2018/assets/estudio-estrategico-nuez-pecanera--2018.pdf**

In terms of value generation by Mexican exports, the pecan nut is the twelfth largest agri-food product, only beer, avocado, tomato, berries, tequila, chili, beef, sugar, cattle, confectionery and chocolate are products that generate more value for their exports than the pecan nut. The value of pecan nut exports exceeds agri-food products such as cucumber, pork, lemons, onions, squash, watermelon and shrimp, and exceeds even other important crops in the trade balance, such as corn and wheat (see Figure 10).

Apple and pecan nut production, like that of many other crops, is classified depending on the production system. Generally speaking, this production system allows production to be classified as High Technology Use (hereinafter AT), Medium Technology Use (MT) and Low Technology Use (hereinafter

BT), precisely according to the technological stock used in production.

Table 3 contains the figures showing the degree of participation in the harvested area, production and production value of MT apple and MT walnut.

share in harvested area, production and
value of production of MT apple and MT walnut, i.e., those
walnut orchards and those apple orchards with MT apple
orchards, i.e., those apple orchards and those walnut orchards
with MT walnut orchards.
i.e. those walnut orchards and those apple orchards
characterised by
characterised by having an average stock in terms of
technology, which, although
technology, which although they are not the most technified
orchards, the most technified ones, are the ones with the
highest production of MT apples.

37

AT farms are also not among the most technologically backward or traditional farms, i.e. those farms with a low stock of technology used in production, i.e. BT farms.

**Table 3. Contribution of apples and walnuts to the agricultural economy of the state of Chihuahua in 2018.**

| Level of aggregation | Surface harvested(ha) | | Production (ton) | | PBV (MX$ nominal 2018) | |
|---|---|---|---|---|---|---|
| The entire state of Chihuahua | 1,004,806.81 | 100% | 13,128,669.82 | 100% | $47,187,481,698.76 | 100% |
| Apple (BT+MT+AT) | 25,244.92 | 2.5% | 569,580.32 | 4.30% | $7,234,514,201.80 | 15.3% |
| Walnut (BT+MT+AT) | 56,951.88 | 5.7% | 101,506.16 | 0.8% | $8,417,000,582.68 | 17.8% |
| Both | 82,196.80 | 8.2% | 671,086.48 | 5.1% | $15,651,514,784.48 | 33.2% |
| The entire state of Chihuahua | 1,004,806.81 | 100% | 13,128,669.82 | 100% | $47,187,481,698.76 | 100% |
| Apple MT | 16,600.01 | 1.70% | 340,860.55 | 2.6% | $4,099,193,109.95 | 8.7% |
| Walnut MT | 18,776.84 | 1.90% | 31,280.36 | 0.2% | $2,571,012,456.09 | 5.4% |
| Both | 35,376.85 | 3.5% | 372,140.91 | 2.8% | $6,670,205,566.04 | 14.1% |

**Source:** Own elaboration based on figures from SIAP "Cierre agricola 2018".

Ash from table 3 it can be seen that in the entire state of Chihuahua, in 2018, a total of 1,004,806.81 ha were harvested, which produced a physical volume of 13,128,669.82 tons, a production volume that had a market value of $47,187,481,608.76, and in relation to these three state totals, the apple and walnut produced by the three levels of technology (BT+MT+AT) occupied 82,196.80 ha (25,244.92 ha of apple and 56,951.88 ha of walnut) equivalent to only 8.2% of the area harvested in the whole state, but in terms of the value of the joint production, apple and walnut (BT+MT+AT) represented 33.2% of the value of the state production, i.e., with only 1 out of every 12 ha harvested in the state, apple and walnut generated 1 out of every 3 pesos of the agricultural value produced in the whole state of Chihuahua.

The lower part of Table 3 indicates only the apple and walnut crops analysed in this water productivity study, i.e. those produced under medium use of MT technology, and their importance in the state's agricultural activity. Table 3 shows that in relation to the total area harvested in the state (just over one million hectares) both crops under MT conditions, with 35,376.85 ha (16,600.01 ha of MT apple and 18,776.84 ha of MT walnut) represented only 3.5% of the area harvested in the state, but their economic importance is enormous, as they contributed 14.1% of the value of the state's agricultural production.

Apples are one of the most expensive items for Mexican families. Mexico is the thirteenth largest apple producer in the world. Despite significant increases in its average national productivity in the period 2003-2016, equivalent to 44.65%, the production of 716,931 tonnes in 2016 covered only 77.26% of the national consumption, so this product is imported fresh, mainly from the United States (97.82% of total imports). In the productive context, of the 58,528 hectares planted in 2016, 79.04% of the area is mechanised, 73.47% has technology applied to plant health, and 61.98% of the territory planted with this crop had technical assistance. On the other hand, 90.24% of the production is under general irrigation and the rest under rainfed irrigation[18] .

## 3.3. Current status of water use efficiency in the production of some fruit trees.

### production of some fruit trees

Information on the productivity and efficiency of water used in the production of some fruit trees is shown in a summarised, ordered and schematic form in table 4. The physical water productivity (WFP), measured in the units of kilograms of apples or nuts produced per cubic metre of water used in production, abbreviated as kg m$^{-3}$ , is meaningful in any comparison, even those that apparently should not be compared, for example the comparison of apple AFP with bovine milk AFP, becomes easier to understand when contrasting the same crop, but produced at different levels of technology use or production system, or in different geographical locations. Ash from table 4 shows that globally, according to Mekonnen and Hoekstra (2011)[19] , 822 L kg-1 are needed, which is a water efficiency index (WEE), which when converted by us to a PFA type index, is suggesting that the use of one m3 of water in apple production at a global average level produces 1,216 kg of apples.

**Table 4. Productivity y physical (PFA, EFA), economic (PEA, EEA) y social (PSA, ESA) efficiency of water used in the production of various fruit trees.**

| Product | site | PFA (kg m' )$^3$ | EAP (USD gain hm' )$^3$ | PSA (Jobs hm' )$^3$ | EFA (L kg' )$^1$ | EEA (m$^3$ /USD gain) | ESA (m$^3$ /employment) | Author |
|---|---|---|---|---|---|---|---|---|
| **Apple BT** | Cuauhtemoc, Chihuahua | 0.91 | $84,341 | 28.1 | 1100 | 11.857 | 35,587 | Rios, Torres y Azpilcueta, 2017 |
| **Apple AT** | Cuauhtemoc, Chihuahua | 3.18 | $614,244 | 22.7 | 314 | 1.628 | 44,053 | Rios, Torres y Azpilcueta, 2017 |

[18] SAGARPA, 2017. Planeacion Agncola Nacional 2017-20130. Manzana Mexicana. Available at: https://www.gob.mx/agricultura/acciones-y-programas/planeacion-agricola-nacional-2017-2030-126813
Mekonnen & Hoekstra, 2011.

|  |  |  |  |  |  |  |  |  |
|---|---|---|---|---|---|---|---|---|
| **Apple MT** | Cuauhtemoc, Chihuahua | 2.02 | $284,726 | 19.6 | 495 | 3.512 | 51,020 | Rios, Torres y Azpilcueta, 2017 |
| **Apple BT** | Canatlan,, Dgo. DDR | 0.88 | $148,405 | 27.9 | 1140 | 6.738 | 35,842 | Rios et al, 2015 |
| **Apple AT** | Canatlan,, Dgo. DDR | 1.88 | $362,825 | 38.6 | 530 | 2.756 | 25,907 | Rios et al, 2015 |
| **Apple** | Santiago Papasquiaro, Dgo | 0.52 | $24,117 | 26.9 | 1910 | 41.465 | 37,175 | Navarrete et al, 2017 |
| **Pecan walnut average B y G** | Delicias, Chihuahua | 0.125 | $52,359 | 3.9 |  | 0.987 | 254,065 | Rios, Torres y Torres,2016 |
| **Pecan walnut average in B y G** | Southwest Coahuila | 0.06 | $10,921 | 17 | 15730 | 91.567 | 58,824 | Rios y Navarrete, 2017 |
| **Pecan walnut** | Torreon, Coahuila | 0.064 | $29,785 | 17.1 | 15560 | 0.000 | 58,318 | Rios, Ruiz y Rios, 2018. |
| **Datil** | Comondu, BCS | 0.103 | $83,000 | 3.84 | 9690 | 11.991 | 260,398 | Zamora y Rivas, 2019 |
| **Olive tree** | Spain | 0.39 | $1,065,739 |  |  |  |  | Montesinos et al (2011) |
| **Apple** | World average |  | $- |  | 822 |  |  | Mekonnen y Hoekstra, 2011 |
| **State production in general** | California, USA |  | $250,000 |  |  |  |  | Fulton, Cooley and Gleick. 2012. |

**Source:** Own elaboration. B = pump-irrigated; G = gravity-irrigated; PEA for pecan walnut from Delicias, Chihuahua y the Olive from Spain, were MX$ 1.01 $m^3$ , y 0.97 € profit per $m^3$ , were deflated with price indices from the bank of Mexico y subsequently standardized to USD (constant 2019) profit per $hm^3$ y the California PEA was USD 0.25 nT .

In the same crop, the apple from Cuauhtemoc, Chihuahua, and moreover, the MT apple, the same apple that will be analysed in this study, according to Rios, Torres and Azpilcueta (2017)[20] had a PFA index of 2.02 kg $_{m-3}$, which in relation to the world average of 1.216 kg $m^{-3}$ , leaves the MT apple from Cuauhtemoc in a good position, since it is 66% more productive in water use than the world average.

The Cuauhtemoc apple produced in BT and AT conditions, according to Rios, Torres and Azpilcueta (2017, *Op. Cit.)* had

---

[20] **Rios-Flores, J.L., Torres M. M. and Azpilcueta RE, M. 2017.** Water productivity in apple trees produced at different levels of technification in Cuauhtemoc, Chihuahua, Mexico. Economic and Administrative Affairs Journal No. 32, first semester 2017. ISSN 0124- 1133, Universidad de Manizales, Colombia.pp.135-146

different PFA indices than the MT apple, since the former had an index of 0.91 kg m-3, while in the latter it was 3.18 kg m$^{-3}$ , higher indices than the corresponding apples from the state of Durango, third producer at national level, where BT and AT apples from Canatlan had indicators of 0.88 and 1.88 kg m$^{-3}$ respectively, while the apple from Santiago Papasquiaro, Durango had a lower PFA (determined by Navarrete *et al, 2017)*[21] , equal to 0.52 kg m$^{-3}$ .

Regarding pecan nut, Rios *et al* (2016)[20] , for pecan nut from Delicias, Chihuahua, on average, without disaggregating by irrigation type (pumping or gravity), i.e., in general, determined a PFA of 0.125 kg m$^{-3}$ , while the similar walnut in general (without disaggregating it by type of irrigation) from southwestern Coahuila, Rios and Navarrete (2017)[22] [23] determined that the PFA of that walnut was 0.06 kg m$^{-3}$ , while the walnut from the municipality of Torreon, according to Rios, Torres and Rios (2018)[24] had a PFA index in the order of 0.064 kg m$^{-3}$ .

According to Hoekstra and Hung (2005)[25] , there are differences between countries in terms of water efficiency in walnut production, for example while Mexico uses 2.811 m3 of water per tonne of walnut, in other major walnut producing countries, Australia uses 2,623 m$^{3}$ ton$^{-1}$ , Argentina 1.702 m$^{3}$


[21] **Navarrete, M, Cayetano, Azpilcueta RE, M., Rios F.J. L. and Torres-M, M. Antonio. 2017.** Agricultural water productivity in apple (*Malus domestica Borkh*) crop produced in the municipalities of Santiago Papasquiaro and Canatlan Durango, Mexico. 2017. In the book: Social Sciences: Economics and Humanities. Macroeconomic variables in agricultural production. Compilers: Perez-Soto, Francisco, Figueroa-Hernandes, Esther, Godinez-Montoya, Lucila and Salazar-Moreno, Raquel. ISBN 978-607-8534-29-6. Pp.93-106. Universidad Autonoma Chapingo, Mexico. Available at: http://www.ecorfan.org/handbooks/

[22] **Rios-Flores, J. L., Torres M. M., Torres M. M. A. 2016.** Agricultural water productivity of pecan nut trees in northern Mexico. Cases: Comarca Lagunera and Delicias, Chihuahua. Editorial Academica Espanola. ISBN 978-3-639-80166-8, Saarbrucken, Germany.

[23] **Rios-Flores, J: L., Navarrete, M. C. 2017.** The metal footprint and economic productivity of water in pecan walnut (Carya illinoensis) in southwestern Coahuila, Mexico. Journal: Estudios de Economia Aplicada. Volume 35-3, September 2017. ISSN 1133-3197. International Association of Applied Economics (ASEPELT), Spain.

[24] **Rios-Flores, J. L., Ruiz Torres, Jose and Rios-Arredondo, Becky Elizabeth. 2018.** Economic-social productivity of water in walnut (*Carya illinoensis*). Case study: Walnut production in the municipality of Torreon, Coahuila, Mexico. Editorial Academica Espanola. ISBN 978-620-2- 13967-0. Beau Bassin, Mauritius

[25] **HOEKSTRA, A. Y. and HUNG, P.Q. (2005).** "Globalization of water resources: international virtual water flows in relation to crop trade". Global Environmental Change, 15, pp. 4556.

ton$^{-1}$ , South Africa 2.759 m3 ton$^{-1}$ , Peru 2.077 m3 ton$^{-1}$ , Israel 949 m3 ton$^{'1}$ , Brazil 2.087 m$^3$ ton$^{-1}$ , Egypt 2.122 m$^3$ ton$^{-1}$ , United States 1.150 m$^3$ ton$^{-1}$ .

According to Zamora and Rivas (2019)[26] determined the PFA of date from Comondu, BCS, equal to 0.103 kg m$^{-3}$ , while the Spanish olive, according to Montesinos *et al* (2011)[27] had a PFA of 0.29 kg m$^{-3}$ .

The inverse of the PFA, the EFA, measured in litres of water per kg of product, abbreviated as L kg$^{-1}$ , is shown in table 4. It is observed that the different apples mentioned in the preceding paragraphs ranged from 314 L kg$^{-1}$ as minimum (At apple from Cuauhtemoc, Chihuahua), to 1,910 L kg$^{-1}$ as maximum (the apple from Santiago Papasquiaro, Durango), being 822 litres the world average.

The ADP column of Table 4 shows that the apple produced at different levels of technification and in the different locations indicated, had very differentiated economic water productivity indices, ranging from USD 24,117 profit per hm$^3$ as a minimum (in the apple of Santiago Papasquiaro, Durango[28] ), to USD 614,244 profit per hm$^3$ as a maximum (in the AT apple of Cuauhtemoc, Chihuahua)[29] . Seen as its inverse, the previous indicator becomes the FSS index, which indicates how much water was necessary to use in production to produce one dollar of profit, as the different blocks ranged in their FSS index from a minimum of 1.628 m3 per USD of profit in the case of the AT block in Cuauhtemoc, to a maximum of 41.465 m$^3$ per USD of profit in the case of the block in Santiago Papasquiaro, Durango.

In the same sense, the pecan walnut presents ADP indices

---

[26] **Zamora, Ramirez, Anselmo & Rivas, L., Jose Antonio. 2019.** Physical, economic and social productivity of water used in the production of datil (*Phoenix dactylifera L.*) irrigated by gravity in Comondu, Baja california Sur. Professional thesis. Universidad Autonoma Chapingo, Unidad Regional Universitaria de Zonas Aridas. Bermejillo, Durango, Mexico.

[27] **Montesinos, P.; Camacho, E.; Campos, B.; Rodriguez_Diaz. J. 2011.** Analysis of Virtual Irrigation Water. Aplication to Water Resources Management in a Meditterranean River Basin. Water Resources management. 25(6): 1635-1651.

[28] Navarrete *et al,* 2017. *Op. cit.*

[29] Rios, Torres and Azpilcueta, 2017. *Op. cit.*

ranging from a minimum of USD 10,921 profit per hm3 in southwest Coahuila[30] , to a maximum of USD 52,359 profit per $hm^3$ in Delicias, Chihuahua[31] , in between is the walnut tree with USD 29,785 $hm^{-3}$ in Torreon, Coahuila[32] .

The PEA, in other fruit trees, such as date from Baja California (Zamora and Rivas, 2019, *Op. Cit.*), and olives from Spain (Montesinos *et al* , 2011, *Op. Cit.*), have PEA indices in the order of USD 83,000 profit per $hm^3$ in the case of date and 0.97 € $m^{-3}$ , equivalent to USD 1,065,739 per $hm^3$ in the case of Spanish olives.

At the level of production in general, not just agriculture, the state of California, according to Fulton, Cooley and Gleick (2012)[33] , is characterised by an index of USD 0.25 value of production per $m^3$ of water, equivalent to USD 250 value per $hm^3$ .

The last but not least form of water productivity, WSP, in Table 4 shows that among the apple cultivars listed at the top, the number of jobs associated with the use of one $hm^3$ of water used in production ranged from a minimum of 19.6 jobs $hm^{-3}$ in the MT apple from Cuauhtemoc, Chihuahua[34] to 38.6 jobs $hm^{-3}$ as maximum in the AT block produced in Canatlan, Durango[35] , intermediate in their PES index were the AT block of Cuauhtemoc, Chihuahua (with 22.7 jobs $hm^{-3}$ ), the BT block of Canatlan, Durango (with 27.9 jobs hm )[-336] , and the block of Santiago Papasquiaro[37] with 26.9 jobs $hm^{-3}$ .

The previous paragraph, seen now in its form as an index of

---

[30] Rios and Navarrete, 2017. *Op. cit.*
[31] Rios, Torres and Torres, 2016. *Op. cit.*
[32] Rios, Ruiz and Rios, 2018. *Op. cit.*
[33] **Fulton, Julian; Cooley, Heather and Gleick, Peter H. 2012. California's Water Footprint.** Pacific institute ISBN 1-893790-46-0 and ISBN 13-978-1-893790-46-9. Oakland, California. Available

in

:

https://indicators.ucdavis.edu/water/files/California%20Water%20Footprint%202012%20Pacific%2 0Institute%20Fulton%20et%20al.pdf . last accessed: February 9, 2020.
[34] Rios, Torres and Azpilcueta, 2017. *Op. cit.*
[35] Rios *et al, 2015, op. cit.*
[36] Rios *et al, 2015. Op. cit.*
[37] Navarrete *et al, 2017.* Op. cit.

efficiency in the use of water used in production, that is, the inverse of the water productivity index, appears in the last column of Table 4, from which it can be seen that the amount of water used in agricultural production ranged from a minimum of 25,907 m3 in the AT block of Canatlan, Durango,[3] to a maximum of 51,020 m3 in the AT block of Cuauhtemoc, Chihuahua, Chihuahua, ranged from a minimum of 25,907 m3 in the AT block of Canatlan, Durango,[38] to a maximum of 51,020 m3 in the AT block of Cuauhtemoc, Chihuahua[39] , with the remaining national blocks being intermediate to these extremes of the SWE index.

The different level of technification is noticeable in the PES indicators for walnut cultivation, since the number of jobs associated with the use of one hm$^3$ of water used in production ranged from 3.9 jobs hm-3 as a minimum in the average pecan walnut tree in Delicias, Chihuahua[40] , one of the most highly technified regions in walnut production in Mexico, that is, without disaggregating the type of irrigation that gave rise to it, up to 17.1 jobs hm$^{-3}$ as maximum in the pecan nut produced in the municipality of Torreon, Coahuila[41] , the walnut tree in the southwest of Coahuila[42] , with 17.0 jobs hm$^{-3}$ is intermediate to these extreme PES rates.

---


[38] Rios, Torres and Azpilcueta, 2017. *Op. cit.*
[39] Rios *et al,* 2015. *Op. cit.*
[40] Rios, Torres and Torres, 2016. *Op. cit.*
[41] Rios, Torres and Rios, 2018. *Op. cit.*
[42] Rios and Navarrete, 2017. *Op. cit.*

# *IV.*MATERIALS AND METHODS
## IV.1. Location of the study area

The free and sovereign state of Chihuahua is one of the thirty-one states, which together with Mexico City, form the thirty-two states of the Mexican Republic. With a surface area of 247,455 $km^2$ , it is the largest state in territorial terms, but with 13.77 inhabitants per $km^2$ it is the third least densely populated state, ahead of Durango and Baja California Sur, the latter being the least densely populated state.

To the north it borders the US states of Texas and New Mexico, to the south with the state of Durango, to the east with the state of Coahuila and to the west mostly with the state of Sonora and a smaller portion of its border in that part is with the state of Sinaloa (see figure 11).

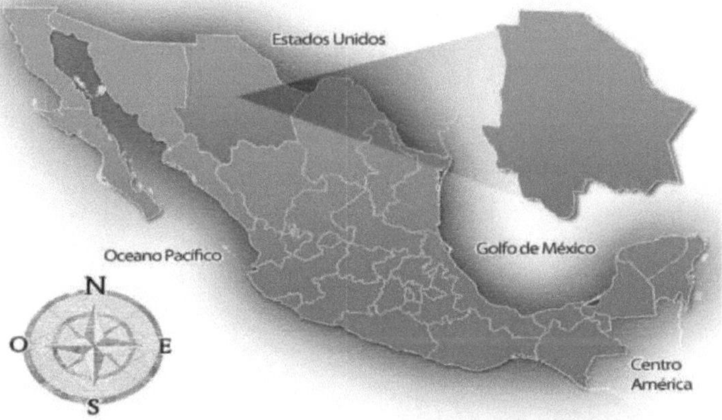


**Figure 11. Location of the state of Chihuahua in the United Mexican States.**

The extreme geographic coordinates of the state are: North 31° 48', South 25° 38' North latitude: East 103° 18', West 109° 07'.

## IV.2. Mathematical models used and the sources from which they were fed.
### were fed

In this work, the mathematical models of Rios *et al* (2015[43] ,

---

[43] **Rios-Flores, J. Luis, Torres M., Miriam, Castro F., Rafael, Torres M., M.A. Ruiz T. Jose. 2015.** Determination of the blue Mdrica footprint in forage crops of DR017 Comarca Lagunera, Mexico. Rev.

$2017^{44\ 45}$ and [2018][43]) shown in Table 5 were used to determine the number of indices of physical, economic and social productivity of water used in production (PFA, PEA and PSA respectively), as well as the indices of physical, economic and social efficiency of water used in production (EFA, EEA and ESA respectively).

## Table 5. Mathematical models used for the edition of physical (PFA), economic (PEA) and social (PSA) productivity of water used in production

FCA UNCUYO, 2018. 47(1): 101-122, ISSN print 0370-4661. ISSN (online) 1853-8665, pp.93-107. Mendoza, Argentina.

[44] **Rios-Flores, J.Luis and Navarrete_Molina, C. (2017).** The monetary footprint and economic productivity of water in pecan walnut (*Carya illinoensis*) in the southwest of Coahuila, Mexico. Journal: Estudios de Economia Aplicada. Volume 35-3, September 2017. ISSN 1133-3197. International Association of Applied Economics (ASEPELT), Valencia, Spain.

[45] **Rfos-Flores, Jose Luis, Rios Arredondo, Becky Elizabeth, Cantu Brito, Jesus Enrique, Rios Arredondo, Hebrian Efrain, Armendariz Erives, Sigifredo, Chavez Rivero, Jose Antonio, Navarrete Molina, Cayetano & Castro Franco, Rafael (2018).** Analisis de la eficiencia ffsica, economica y social del agua en esparrago (*Asparagus officinalis L.*) y uva (*Vitis* vinifera) mesa del DR-037 Altar-Pitiquito-Caborca, Sonora, Mexico 2018. *Revista de la Facultad de Ciencias Agrarias. National University of Cuyo*, 50(2). ISSN in print 0370-4661, ISSN (online) 1853-8665. Mendoza, Argentina.

| Variable | Modelo para un cultivo en lo individual | Modelo para un agregado grupal cultivos |
|---|---|---|
| 1) PFA (en L kg$^{-1}$) | $Y = 10^{4}\,LRi\,(RFi\,ECi)^{-1}$ | $y = \dfrac{10^{4}\,\sum\limits_{i=1}^{n} S_i\;LR_i\;(EC_i)^{-1}}{\sum\limits_{i=1}^{n} S_i\;RF_i}$ |
| 2) EFA (kg m$^3$) | $Y = 10^{-1}\,RFi\,ECi\,LRi^{-1}$ | $y = \dfrac{10^{-1}\,\sum\limits_{i=1}^{n} S_i\;RF_i}{\sum\limits_{i=1}^{n} S_i\;LR_i\;(EC_i)^{-1}}$ |
| 3) EEA (m$^3$ USD de ganancia$^{-1}$) | $y = \dfrac{10^{4}\left(\dfrac{LR_i}{EC_i}\right)}{RF_i\left(\dfrac{P_i}{PC}\right)-\left(\dfrac{C_i}{PC}\right)}$ | $y = \dfrac{10^{4}\,\sum\limits_{i=1}^{n} S_i\;LR_i\;(EC_i)^{-1}}{\sum\limits_{i=1}^{n} S_i\;((RF_i\;P_i - C_i)/PC)}$ |
| 4) PEA (miles de USD de ganancia hm$^{-3}$) | $y = 10^{2}\,g_i\,EC_i\,(LR_i)^{-1}$ | $y = \dfrac{10^{2}\,\sum\limits_{i=1}^{n} S_i\;g_i}{\sum\limits_{i=1}^{n} S_i\;LR_i\;(EC_i)^{-1}}$ |
| 5) PSA (Empleos hm$^{-3}$) | $y = \dfrac{25\;J_i}{72\;(LR_i/EC_i)}$ | $y = \dfrac{25\;\sum\limits_{i=1}^{n} S_i\,J_i}{72\;\sum\limits_{i=1}^{n} S_i\,(LR_i/EC_i)}$ |
| 6) ESA (m$^3$ empleo$^{-1}$) | $y = \dfrac{2.88 \ast 10^{6}\,LR_i}{J_i\,EC_i}$ | $y = \left(2.88 \ast 10^{6}\right)\dfrac{\sum\limits_{i=1}^{n} S_i (LR_i/EC_i)}{\sum\limits_{i=1}^{n} S_i J_i}$ |



The meaning of each of the independent variables on which the PFA, PEA, PSA, EFA, EEA and ESA depend, as well as the source from which they were taken, are as follows:

RFi = Physical yield of the i-th crop (in ton ha$^{-1}$ ). Where RF = VBP/P; the sources of VBP (Gross Value of Production and S = harvested area (in ha) are from SIAP-SADER (2020). "Cierre agncola 2018". Available at https://nube.siap.gob.mx/cierreagricola/

LRi = Irrigation line of the i-th crop (in m). Source: INIFAP-INIFAP-CENID-RASPA, 2006[46] .

---

[46] **INIFAP-CENID-RASPA. (2006).** *Irrigation Programme.* [Accessed on: 1 September 2006 2019]. Available at: https://cenidraspa.org/serg/serg_v1.php

ECi = Efficiency of hydraulic conduction of the i-th crop. 0 < EC < 1 (Source: INIFAP.CENID-RASPA, 2006 *Op. Cit.*).

Si = Harvested area of the i-th crop (in ha). Source: SIAP-SADER (2019). Source: SIAP-SADER (2018 *Op. Cit.*).

pi = Price of the product of the i-th crop (in MX$ ton$^{-1}$ ). Where pi = VBPi/Pi, the sources of VBP (Gross Value of Production) and P = annual physical production (in tonnes) are from SIAP-SADER (2018 *Op. Cit.*).

CP = Exchange rate, Mexican pesos (MX$) per USD. Source: Currency converter: available at: https://www.xe.com/es/currencyconverter/convert/?Amount =1&From=USD&To=MXN

Ci = Cost of production per hectare of the i-th crop (in MX$$^{-1}$ ). Source: FIRA (2018, "Agrocost", available at: https://www.fira.gob.mx/Nd/Agrocostos.jsp)

gi = Ui = Profit per hectare of the i-th crop (in US$ ha$^{-1}$ ) = RFi(pi/PC)-(Ci/PC).

Ji = Number of days invested per hectare in the i-th crop. Sources: FIRA (2018, "Agrocost", available at: https://www.fira.gob.mx/Nd/Agrocostos.jsp).

i = i-th crop under a particular form of irrigation (pumping, gravity, drip irrigation, sprinkler irrigation, etc.).

288 = Number of working days per year per worker = 6 working days per week, times 48 weeks per year.

## IV.3. Geographical delimitation, and meaning of abbreviations

The data on harvested area, production and value of production for MT pecan nut and MT apple, shown in table 7, are from the DDRs, CADERs and municipalities shown in Annex 1.

Meaning of the following abbreviations used in this work:

LV, MV and HV: literally mean Low Technology Use, Medium Technology Use and High Technology Use.

DR: Irrigation District

RDD: Rural Development District

CADER: Support Centre for Rural Development

m3: cubic metre

hm3: cubic hectometre = one million cubic meters

USD: US dollar

MUSD: Million US Dollars

MX$: Mexican Peso

# *V.* RESULTS AND DISCUSSION.
## V.1. Profitability and productivity indicators and efficiency of water used in MT apple and MT walnut production in Chihuahua.

Table 6 records the production costs per hectare for both MT walnut and MT apple on average for the whole state of Chihuahua. From this it can be seen that the total cost is composed of two types of costs: operating costs and other costs. Operating costs in turn are broken down into land preparation, fertilisation, cultural work, pest and weed control, irrigation, harvesting and miscellaneous, while "other costs" are broken down into land rent, financial cost and depreciation of machinery and equipment.

The total cost of each crop was MX$ 139,376/ha (equivalent to USD 7,225 per ha) in the case of MT apple and MX$ 109,998 (equivalent to USD 5,702 per ha) in the case of MT walnut. The percentage weight of operating costs and other costs was different in each crop, as in MT apple, of the USD 7,225 per ha, 70.5% (USD 5,096) was in operating costs and 29.5% (USD 2,129) in other costs, while in the case of MT walnut, of the USD 5,702 per ha total cost, 65.4% (USD 3,727) was in operating costs and 34.6% (USD 1,975) in other costs (see table 6). The total cost per ha in apple MT was 79% higher than in walnut MT.

The percentage weight of each item within the total cost per hectare, differs in each crop, so^ for example, very particularly for the irrigation item, given the nature of this work on productivity and efficiency of water used in production, it is observed in table 6 that with MX$ 9,072 and MX$ 16,115 respectively for MT apple and MT walnut, the cost of water used in production represented 6.5 and 14.7% respectively, suggesting that in absolute terms, the MT walnut producer spent 78% more on water used in production per hectare than the MT apple producer.

**Table 6. Production costs per hectare in pecan and apple trees produced under medium technology use (MT) conditions in Chihuahua, Mexico, 2018.**

| Type of cost | Apple MT | | Pecan walnut MT | |
|---|---|---|---|---|
| Operating costs | Cost (MX$/ha) | Daily wages | Cost (MX$/ha) | Daily wages |
| Land preparation | | | $2,394 | |
| Fertilisation | $11,204 | 9 | $17,320 | |
| Cultural work | $57,564 | 33 | $8,904 | 17 |
| Pest and weed control | $13,108 | | $9,706 | |
| Irrigation | $9,072 | 0.9 | $16,115 | 11 |
| Harvest | $5,914 | 18 | $14,160 | |
| Various | $1,440 | | $3,300 | |
| **Subtotal operating costs** | **$98,302** | | **$71,899** | |
| **Subtotal operating costs** | **$5,096** | | **$3,727** | |
| Other costs: | | | | |
| Land rent | $30,000 | | $30,000 | |
| Financial cost | $9,108 | | $6,661 | |
| Depreciation of machinery and equipment | $1,966 | | $1,438 | |
| **Subtotal other costs** | **$41,074** | | **$38,099** | |
| **Subtotal other costs** | **$2,129** | | **$1,975** | |
| **TOTAL COST (MX$)/ha** | **$139,376** | **60.9** | **$109,998** | **28** |
| **TOTAL COST (USD)/ha** | **$7,225** | | **$5,702** | |
| m /ha$^3$ | 9000 | | 11000 | |
| Hours/ha | | 487.2 | | 224 |
| Cost of water (MX$/m )$^3$ | $1.008 | | $1.465 | |
| Cost of water (USD/m )$^3$ | $0.052 | | $0.076 | |
| peso-USD exchange rate | 19.29 | | | |

**Source:** Own elaboration, the operating costs are sourced from the production costs of AGROFIRME SA de CV SOFOM ENR and the "other costs" of land rent and financial cost from the production costs of FIRA in its page of AGROCOSTOS, depreciation was taken based on the Prontuario Fiscal 2019, at the rate of 2.5% of operating costs.

Percentage-wise, according to table 6, the most expensive cost was cultural work with 41% of the total cost per hectare, followed far behind, in second place, by fertilisation with 8%, while in MT walnut the most expensive was fertilisation with 16% of the total cost, and in second place was irrigation with 14.7% of the total cost per ha.

It should be mentioned that FIRA's production costs per hectare, in its "Agrocost" portal, are only composed of operating costs, and do not include land rent, financial costs or depreciation of machinery and equipment, so we considered it necessary, in order to determine a profitability as close to reality as possible, to add these other costs, which were precisely called "other costs" in table 6: "other costs", as otherwise an artificially high profitability (measured by the Profit/Cost Ratio) would have been obtained, and thus not reflecting the producer's reality.

Table 6, from our perspective, determines the total cost per hectare from three perspectives: the monetary one, which has already been discussed, the one that talks about the investment of labour time per hectare and the one that suggests a third type of cost, the "water cost", i.e. how much water was necessary to use at commercial scale in order to obtain the output quantity of physical product, which when multiplied by the price becomes the income per hectare, which in turn, if the monetary cost is subtracted, yields the monetary profit per hectare.

Thus, from the perspective of how much labour was necessary to invest per hectare, table 6 shows that 60.9 working days (it is assumed that a working day is 8 hours of labour) were necessary in the MT apple and 28 working days in the MT walnut, which in terms of working hours is equivalent to 487.2 hours of average social work invested per hectare in MT apple and 224 hours of average social work invested in MT walnut.

In the same sense, and from the perspective of "water cost", table 6 shows that 9,000 m3 of water were used per hectare in MT apple and 11,000 $m^3$ of water per hectare in MT walnut, so that, when dividing the irrigation item by this volume of water, we obtain the indicator that estimates the cost per m3 of water used in the production indicated in the lower part of table 6, table that shows that to the MT apple producer it cost on average USD 0.052 per $m^3$ of water (equivalent to MX$ 1.008 $m^{-3}$ ) while the same volume of water cost the walnut producer

USD 0.076 (equivalent to MX$ 1.465), 45% more expensive than the MT apple producer.

Table 8 shows the figures for production, profitability and water and labour use for MT walnut and MT apple crops in the state of Chihuahua in 2018.

**Table 7. Production, profitability, water use and labour time in the production of MT Apple and MT Pecan Walnut crops in the state of Chihuahua, Mexico, 2018.**

| Variable | Apple MT | Walnut MT | Both crops | Walnut/ Apple |
|---|---|---|---|---|
| Harvested area (ha) | 16,600.01 | 18,776.84 | 35,376.85 | 1.13 |
| Annual production (ton) | 340,860.55 | 31,280.36 | 372,140.91 | 0.09 |
| VBP (MMX$) | $4,099.19 | $2,571.01 | $6,670.21 | 0.63 |
| VBP (THUS$) | $212.50 | $133.28 | $345.79 | 0.63 |
| Yield (ton)/ha | 20.534 | 1.666 | 10.519 | 0.08 |
| PMR (MX$/tonne) | $12,026 | $82,193 | $17,924 | 6.83 |
| PMR (USD/tonne) | $623 | $4,261 | $929 | 6.83 |
| Income (MX$)/ha | $246,939.20 | $136,924.66 | $188,547.19 | 0.55 |
| Income (uSD)/ha | $12,801.41 | $7,098.22 | $9,774.35 | 0.55 |
| Operating costs (MX$)/ha | $98,302.00 | $71,899.00 | $84,288.18 | 0.73 |
| Operating costs (USD)/ha | $5,096 | $3,727 | $4,370 | 0.73 |
| Operating profit (MX$/ha) | $148,637.20 | $65,025.66 | $104,259.01 | 0.44 |
| Operating profit (USD/ha) | $7,705.40 | $3,370.95 | $5,404.82 | 0.44 |
| Total cost (MX$)/ha | $139,376 | $109,998 | 123,783.24 | 0.79 |
| Total cost (USD)/ha | $7,225 | $5,702 | 6,417 | 0.79 |
| Gross profit (MX$/ha) | $107,563.47 | $26,926.24 | $64,763.95 | 0.25 |
| Gross profit (USD/ha) | $5,576.13 | $1,395.87 | $3,357.38 | 0.25 |
| RB/C with operating costs | 2.51 | 1.90 | 2.24 | 0.76 |
| RB/C at full cost | 1.77 | 1.24 | 1.52 | 0.70 |
| $hm^3$ of water consumed in the entire harvested area | 149.40 | 206.55 | 355.95 | 1.38 |
| Gross profit on all area harvested (MUSD) | $92.56 | $26.21 | $118.77 | 0.28 |
| Working days on the entire harvested area | 1,010,941 | 525,752 | 1,536,692 | 0.52 |
| Equivalent jobs | 3,510 | 1,826 | 5,336 | 0.52 |
| Capital investment on all harvested area (MUSD) | $119.94 | $107.07 | $227.01 | 0.89 |

**Source:** Own elaboration, based on figures from SIAP's "Cierre agncola 2018" and table 6. MX$= Mexican pesos. MUSD = Millions of US dollars.

This source indicates that a total of 35,376.85 ha of both crops were harvested in 2018, producing a combined volume of

372,140.91 tons, which had a market value of MUSD[47] 345.79; once the above figures are disaggregated, it can be seen that the total area indicated is broken down into 18,776.84 ha of MT walnut (53.1%) and 16,600.01 ha (46.9%) of MT apple; of the 362,130.91 tons produced, 340,860.55 tons (91.6%) were apple and 31,280.36 tons (8.4%) were walnut; while of each of the MUSD 345.79 of joint value generated, 61.5% was contributed by apple and 38.5% by walnut. See table 7.

The gross profit per hectare amounted to USD 5,576.13 in the MT apple and USD 1,395.87 in the MT walnut, which translated, given their income per hectare, into a BR/C of 1.77 and 1.24 respectively, which suggests that although both BR/C were good, as both were above unity, it was much better in the MT apple than in the MT walnut, since while in the walnut it was possible to recover each USD invested in production and an additional surplus of USD 0.24 in the form of gross profit before taxes, in the apple the RB/C indicator was much better, as it was possible to recover each USD invested in production, but also obtained a profit equal to USD 0.77; the above, in the form of profit rate, suggests an average profit in the order of 24% in the walnut and in the order of 77% in the apple. See table 7.

All the harvested area of both crops, according to table 7, produced a mass of profit equal to MUSD 118.77, of which MUSD 92.56 (77.9%) was generated by the MT apple and MUSD 26.21 (22.1%) by the MT walnut, now, to produce that mass of profit, both crops used in production a total of 355.95 million $m^3$ of water, that is, 355.95 $hm^3$ , of which, MT apple used 42% (149.40 $hm^3$ ) and walnut the remaining 58% (206.55 $hm^3$ ), also, of the 5,336 jobs that were jointly generated in the production of both fruits, MT apple contributed 65.8% (3,510 jobs) and walnut the remaining 34.2% (1,826 jobs). This suggests that walnut had little macroeconomic efficiency in the use of soil, water and capital resources in production, since with

---

[47] MUSD, initials in millions of US dollars.

53% of the soil used jointly by both crops, 58% of all the water used jointly by both crops and 47% of all the capital invested in the production of both fruits, it generated only 39% of the combined GVP, It also generated only 22% of the combined profit and 34% of the combined employment, while the MT apple behaved efficiently in using scarce soil, water and capital resources, because with 47% of the combined soil used by both crops, 42% of all the water used by both crops and 53% of the combined capital investment (equal to MUSD 227.01) achieved the objectives of generating 61% of the joint GVP, 78% of the joint mass of profits and 66% of all joint employment generated by both crops. See tables 7 and 8.

**Table 8. Macroeconomic efficiency of water, soil and capital resource use in relation to PPV, profits and employment achieved by using such resources in MT apple and MT walnut production in Chihuahua, Mexico, 2018.**

| Uso de recursos: | Manzana MT | Nogal MT | Ambos |
|---|---|---|---|
| Suelo | 47% | 53% | 100% |
| Agua | 42% | 58% | 100% |
| Capital | 53% | 47% | 100% |
| Objetivos logrados: | | | |
| VBP | 61% | 39% | 100% |
| Ganancias | 78% | 22% | 100% |
| Empleo | 66% | 34% | 100% |

**Source:** Own elaboration, based on table 7.

This suggests that water, soil and capital are scarce goods, especially fresh water available for human use, which represents only 0.26% of the total amount of water available for human use, and is only 0.26% of the total amount of water available for human use, which is only 0.26% of the total amount of water available for human use.

of all the water on the planet should be used more efficiently, more productively on walnut trees, which is possible to achieve not only by means of large and difficult changes such as genetic improvement of varieties, or extremely difficult to achieve technological improvements in irrigation, no. small and seemingly insignificant modifications are enough, for example, timely and timely weeding, cleaning of irrigation channels, application of insecticides, fertilisers and agrochemicals, Sometimes small and seemingly insignificant modifications are enough, for example, timely and opportune weeding, cleaning of irrigation canals, application of insecticides, fertilisers and agrochemicals in general at the most favourable times and not when the pests have already passed the economically viable threshold, adequate weeding, etc.

In terms of marginal water productivity, dividing the corresponding contribution percentages, it is obtained that

In the case of MT walnut, for every 1% of the total combined water volume used, MT walnut contributed 0.38% and MT apple 1.86% of the total combined gains, and for every 1% of the total combined water volume used, MT walnut generated 0.59% of the total combined employment, while MT apple generated 1.567% of the total combined employment (table 8).

Table 9 is the most important part of this paper, as that source contains the index numbers of the physical, economic and social productivity and efficiency of water used in MT apple and MT walnut production in Chihuahua, Mexico, in 2018.

### Table 9. Physical, economic and social productivity and efficiency of water used in MT apple and <u>MT walnut production in Chihuahua, Mexico, 2018.</u>

| Variable | Units | a) Apple MT | b) Nut MT | c )= a/b |
|---|---|---|---|---|
| **Efficiency (E) and productivity (P) physical (F) of water (A) used in production** | | | | |
| EFA | Litres of water/kg | 438 | 6,603 | 0.066 |
| PFA | $^3$ kg m$^3$ | 2.28 | 0.15 | 15.065 |
| **Economic (E) efficiency (E) and productivity (P) of water (A) used in the production** | | | | |
| EEA | m$^3$ of water used per USD profit | 1.61 | 7.88 | 0.205 |
| EAP | USD profit per m$^3$ of water used in production | $0.62 | $0.13 | 4.88 |
| **Efficiency (E) and social (S) productivity (P) of water (A) used in production** | | | | |
| ESA | m$^3$ of water used per employment generated | 42,562 | 113,143 | 0.376 |
| PSA | Jobs generated by hm$^3$ | 23 | 9 | 2.66 |
| **Water prices** | USD per m$^3$ | $0.052 | $0.076 | |

Horizontally, table 9 is divided into three parts, the first part contains the indicators of physical efficiency and productivity of water used in production, the central part shows the index numbers of economic efficiency and productivity of water used in production, and the third part refers to the indicators of social efficiency and productivity of water used in production.

In that sense, the upper part of table 9 suggests, according to the EFA (physical water efficiency) indicator, that producing one kg of walnut used 6,603 litres of water on average in Chihuahua in the medium-technology-use walnut orchards, while producing that same kg, but from the apple produced in the medium-technology-use orchards, used only 438 litres of water. This suggests that producing one kg of biomass in the MT apple uses only 6.6% of the volume of water used in the MT walnut to produce the same kg of biomass.

In terms of physical water productivity (PFA), table 9 shows indicators of 0.15 and 2.28 kg per m$^3$ of water in the production of Mt walnut and MT apple respectively, which suggests that the apple produces 15.065 times more biomass per volumetric unit of water than the pecan nut, or seen in another way, that the physical water productivity of the MT walnut is only 6.6% of the physical water productivity of the MT apple.

Regarding the productivity and economic efficiency of the water used in production, Table 9 shows that the use of the same volume of water, one hm$^3$ , produced USD 619,570 (equivalent to USD 0.62 profit per m$^3$ ) in the MT apple and USD 126,897 (equivalent to USD 0.13 profit per m3) in the MT walnut.13 profit per m3) in the MT walnut, thus inferring that the productivity of water, in economic terms, is in the MT apple 4.88 times the level of productivity achieved by the MT walnut, or seen in another way, that the MT walnut water, in economic terms, produces only 20.5% of the profit that the same volume of water produces in the MT apple.

In terms of economic efficiency of water, the FSS indicators in table 9 show that producing one US dollar of profit implied the use of 1.61 $m^3$ in the case of MT apple and 7.88 $m^3$ in the case of MT walnut. This indicates that there is a greater economic efficiency of water in the MT apple crop, since in this crop, to produce one US dollar of profit, only one fifth, i.e. 20.5% of the volume of water used in the MT walnut to produce the same US dollar of profit is used.

In its lower part, table 9 shows that the social productivity of water used in production generated 23 jobs per $hm^3$ in the MT apple and only 9 jobs per $hm^3$ in the MT walnut, that is, in the MT walnut, water is used in a less productive way in social terms, since the same volume of water, one $hm^3$ , produces only 37.6% of the employment that the same volume of water generates in the MT apple, which seen from the perspective of the MT apple suggests that this crop produces 1.66 times (the indicator is 2.66 = 1 + 1.66) more employment than that produced by the MT walnut when using the same volume of water in production.

Producing one job, in the case of the walnut had a "cost" of 113,143 $m^3$ of water, while producing the same job in the MT apple implied an investment of only 42,562 $m^3$ of water, i.e. with the amount of water used in the MT walnut to produce one job, 2.66 jobs were produced in the MT apple, according to table 9.

### 5.2. Discussion

The price of water is extremely important, since it is a good that appears to be infinite, when in reality water is extremely scarce, so that when the price of a good is very low, the producer does not make efficient use of this apparently infinite good. The price of water is a social rather than an economic indicator, although it indicates how much water costs the agricultural producer[48] , and in reality, it is an indicator of an eminently social nature,

---

[48] The water price shown in table 10 is derived by dividing the amount of the irrigation cost item (within the costs per hectare) by the volume of water used per hectare.

since water is an extremely limited resource and, to date, water, according to the Constitution of the United Mexican States, belongs to the whole of society, However, this social resource is used privately by agricultural, livestock and industrial producers, so that in fact a private appropriation of profits is made, which have been generated with a resource that does not belong to the producer, but belongs, at least legally, to society as a whole.

In relation to the price per $m^3$ of water used in production, it should be remembered that this price in Chihuahua amounted to USD 0.076 $m^{-3}$ and USD 0.052 $m^{-3}$ in MT pecan walnut and MT apple respectively, being the price of walnut in Chihuahua very similar to the price of water in another fruit tree, grape, since its price was similar to USD 0.075 and USD 0.070 per $m^3$ for industrial grape and table grape crops produced in Coahuila (according to Carrillo, 2019[49] ) and Caborca, Sonora (according to Rios *et al, 2018, Op. Cit.*[50]) respectively, of all the other crops against which the price of water in the Chihuahua pecan nut is contrasted in Table 10, after the two vines already mentioned, only the MT apple analysed in this paper is the closest, as mentioned above, with a price of USD 0.052 per $m^3$ of water used in production.

The price per cubic metre of water used in agriculture in other agricultural regions of the world, as, for example, Gleick (2000)[51] notes that farmers in the United States, on average, pay a price of US$0.05 $m^{-3}$ for water used in agricultural irrigation, while the US public sector, the author notes, pays an

[49] **Carrillo, C. J. 2019**. Economic-social productivity of water in drip-irrigated grapevine (*Vitis vinifera)* in Coahuila, Mexico. Professional thesis. Unidad Regional Universitaria de Zonas Aridas, Universidad Autonoma Chapingo. Bermejillo, Durango, Mexico.

[50] **Rfos-Flores, Jose Luis, Rios Arredondo, Becky Elizabeth, Cantu Brito, Jesus Enrique, Rios Arredodndo, Hebrian Efrain, Armendariz Erives, Sigifredo, Chavez Rivero, Jose Antonio, Navarrete Molina, Cayetano & Castro Franco, Rafael (2018)**. Analysis of the physical, economic and social efficiency of water in asparagus (*Asparagus officinalis L.)* and grapes (*Vitis* vinifera) table of DR-037 Altar-Pitiquito-Caborca, Sonora, Mexico 2018. *Revista de la Facultad de Ciencias Agrarias. National University of Cuyo*, 50(2). ISSN in print 0370-4661, ISSN (online) 1853-8665. Mendoza, Argentina.

[51] **Gleick (2000)**. The World's Water, 2000-2001: The Biennial Report on Freshwater Resources. Washington, DC. Islan Press, 2000. 335p.

amount ranging from US$0.30 m$^{-3}$ to US$0.80 m$^{-3}$ for water treated for personal use.

The paradox or tragedy of the commons, as pointed out by economic science, shows that when resources are in common use and are not priced at all, they tend not to be used efficiently or productively, hence the price of water, in principle, should exist, secondly, it should be priced at something that obliges the producer who uses that water to use it in the most efficient and productive way, Hence, authors such as Takele and Kallenbach (2001)50 , point out that the price of water is important both to improve the demand for and conservation of this scarce resource, but globally there are also examples that the resource is not valued as a finite resource. In a study covering farmers in the very important Imperial Valley region in the US state of California, a region characterised as the main supplier of vegetables for the whole of the United States of America, Murphy (2003)[52] [53] found that farmers in this agricultural region pay on average only US$15.50 per 1,200 m3 (1,200 m3) for water.50 per 1,200 m3 (i.e. US$ 0.012 m$^{-3}$ ). Likewise, for one of the most important irrigation districts in Mexico, DR017, located in the north of the country, in La Comarca Lagunera, Rfos *et al,* (2015 *op. cit*), determined a weighted average price of water in the main forage crops, of the order of US$0.02 m$^{-3}$ , while, the same author, in another work[54] , and for the region of the Mexicali Valley, determined a weighted average price of irrigation water equal to $0.19 m$^{-3}$ .

Considering some reference for the above figures, for example, the average market price of a litre of bottled water, MX$10 (equivalent to = US$ 541.24 m$^{-3}$ ), indicates that in principle,

---

[52] **Takele, E. and Kallenbach, R. (2001)**. Analysis of the Impact of Alfalfa Forage Production under Summer Water-Limiting Circumstances on Productivity, Agricultural and Growers Returns and Plant Stand. Journal of Agronomy and Crop Science, 187 (1): 41-46.

[53] **Murphy D.E. (2003)**. In a first, U.S. puts limits on California's thirst. New York Times, 5 January. 1-16 p.

54 **Rfos, F. J. L., Torres, M. M., Ruiz, T. J. and Torres, M. M. A. (2016)**. Irrigation water efficiency and productivity in wheat (Triticum vulgare) from Ensenada and Mexicali Valley, Baja California, Mexico. Acta Universitaria 26(1): 20-29.

wherever the agricultural producer is from, the value assigned to the m3 of water used for irrigation is almost tending to a price equal to zero, since in the case of the Imperial Valley of California, in the case of DR017 forage crops and in the case of wheat in the Mexicali Valley, it represents only 0.0022%, 0.0000369% and 0.035% of the price paid by the consumer of bottled water, and this *necessarily* makes the farmer believe that the water resource has no value whatsoever, so that the importance of efficient water use is underestimated, making the most productive use of every drop of water used in agricultural production. This should not be misunderstood, it should not be thought that the farmer should be punished with excessive water prices, no, because food production by the agricultural sector is undoubtedly a *strategic* branch of production that guarantees *food security*, moreover, agricultural production can be assigned the character of *national security*, but this does not mean that agriculture should not be obliged to be efficient in the use of water, to be sustainable in the long term.

**Table 10. Price of water used in the production of some crops**

| Crop/location | Reported by the author | | Author |
|---|---|---|---|
| **Product** | Price/m3 | Units | Work |
| **Nogal MT, Chihuahua** | $0.076 | USD/m3 | This work |
| **Manzana MT, Chihuahua** | $0.052 | USD/m3 | This work |
| **Manzana BT, Canatlan, Dgo.** | 0.02 | USD/m3 | Rios *et al*, 2015 |
| **Manzana AT, Canatlan, Dgo.** | 0.02 | USD/m3 | Rios *et al*, 2015 |
| **Pecan walnut average B and G, Delicias, Chih.** | MX$ 0.55/m$^3$ (=USD 0.037/m3) | | Rios, Torres and Torres,2016 |
| **Pecan walnut average in B and G, Southwest Coah.** | $0.0119 | USD/m3 | Rios and Navarrete, 2017 |
| **Pecan walnut, Torreon, Coah.** | $0.016 | USD/m3 | Rios, Ruiz and Rios, 2018. |
| **Datil** | $0.008 | USD/m3 | Zamora and Rivas, 2019 |
| **Vid, Caborca, Son.** | $0.070 | USD/m3 | Rios *et al, 2018* |
| **Vid, Coahuila** | $0.075 | USD/m3 | Carrillo, 2019 |
| **Avocado, Patzcuaro, Mich.** | $0.012 | USD/m3 | Rios, Ruiz and Azpilcueta, 2018 |

**Source:** Own elaboration. B = pumped irrigation G = gravity irrigation

The price per m3 of water used in the production of the

remaining crops indicated in table 10 is far from the price of water, not only in the MT walnut crop, but also in the MT apple of the state of Chihuahua, since in the BT and AT apple of Canatlan, Durango, the price of water, according to Rios *et al, 2015 (Op. Cit.)* was USD 0.02 m$^{-3}$ in both apples (see table 10). In the same sense, in pecan walnut trees in Chihuahua, southwest Coahuila and Torreon, Coahuila, with USD 0.037 (equivalent to MX$ 0.55)[55] , USD 0.0119[56] and USD 0.016[57] the m$^3$ of water used in production respectively, result in only 48.6%, 16.7% and 21.1% of the price of water paid by the MT pecan walnut farmer in Chihuahua (see table 10).

Finally, in relation to other fruit trees such as date (*Phoenix dactylifera*) produced in Comondu, Baja California Sur[55][58] and avocado (*Persea americana Miller*) from the state of Michoacan[59] , crops in which the average price per cubic metre of water used in production was equal to USD 0.008 and USD 0.012 respectively, are very low water prices in relation to the price of water for MT walnut and MT apple from the state of Chihuahua, however, it should be noted that in the case of date, it is gravity-irrigation water, It is not a price of underground water, much less of any specialised type of irrigation, while in the case of avocado, it is underground water, yes, but of traditional irrigation, that is, once it was extracted from the subsoil, it was irrigated by gravity to the orchard, which is why they are relatively low water prices.

Pedroza *et al,* [201458] , determined that in alfalfa and forage ma^z irrigated with groundwater by traditional pumping in La Laguna, had a water price equal to MX$ 0.42 and MX$ 0.45, prices that

---


[55] Rios, Torres and Torres, 2016. *Op. cit.*

[56] Rios and Navarrete, 2017. *Op. cit.*

[57] Rios, Ruiz and Rios, 2018. *Op. cit.*

[58] **Zamora, R., A. & Rivas, L., J. A. 2019.** Physical, economic and social productivity of water used in the production of datil (*Phoenix dactylifera L.*) irrigated by gravity in Comondu, Baja california Sur. Professional thesis. Universidad Autonoma Chapingo, Unidad Regional Universitaria de Zonas Aridas. Bermejillo, Durango, Mexico.

[59] **Rios-Flores. J. L., Ruis Torres, J., Azpilcueta Ruiz Esparza, M. de J. 2019**. Economic-social productivity of water in avocado cultivation. The case of production in Michoacan, Mexico. Editorial Academico Espanola. ISBN 978-3-639-53183-1. Beau Bassin, Mauritius.

brought to present value in constant 2018 pesos using an inflation rate of 3% annual increase in prices and at the exchange parity considered in this study, equal to MX$ 19.29 per USD, yield USD 0.0252 and USD 0.0270 per m3 respectively, consistent with the water prices reported in Table 10, similar to the MX$ 0.19 m$^{-3}$ (equivalent to USD 0.012 m$^3$ ) reported by Rios *et al* (2016 a)[60] [61] for the price of water in the wheat crop produced in San Luis Rio Colorado, Sonora; among forages in DR017 of La Comarca Lagunera, the average water price, according to Rios *et al,* 2015[62] , was USD 0.02 m$^{-3}$ .

The data on water productivity and efficiency, in physical, economic and social terms, i.e. PFA, PEA and PSA data as well as EFA, EEA and ESA already mentioned in the Literature Review chapter, are presented in schematic and summary form, and are contrasted against the results found in this study for MT walnut and MT apple in the state of Chihuahua, which are presented in table 11.

Thus, table 11 shows that the AFP, for the purpose of comparison and contrast, is shown already standardised to the same units of measurement: kg of physical product produced per m$^3$ of water used in production, or simply abbreviated as kg m$^{-3}$ , which assumes the form of reference in the case of the MT apple crop from Chihuahua, so that its 2.28 kg m$^{-3}$ are considered as the unit against which the other crops from that source are contrasted. Ash table 11 shows that only the AT apple from Cuauhtemoc, Chihuahua[63] had an index (which

---


[60] **Pedroza Sandoval, Aurelio; Rios Flores, Jose Luis; Torres Moreno Myriam; Cantu Brito, Jesus Enrique; Piceno Sagarnaga, Cesar; Yanez Chavez, Luis Gerardo. 2014**. Irrigation water efficiency in the production of forage maize (*Zea mays L.)* and alfalfa (*Medicago sativa)*: Social and economic impact. Terra Latinoamericana, vol. 32, num. 3, July-September, 2014. Pp. 231-239. Sociedad Mexicana de la Ciencia del Suelo, A. C., Chapingo, Mexico.

[61] **Rios-Flores Jose Luis, Torres Moreno Miriam, Ruiz Torres Jose, Torres Moreno Marco Antonio. 2016**. Efficiency and productivity of water irrigation in wheat (*Triticum vulgare*) from Ensenada and Valle de Mexicali, Baja California, Mexico. Acta Universitaria Journal. Multidisciplinary Scientific Journal. ISSN 0188-6266 Vol. 26 No. 1. January-February 2016. Guanajuato, Mexico.

[62] **Rios Flores, Jose Luis; Torres Moreno Miriam; Castro Franco Rafael; Torres Moreno Marco Antonio. 2015.** Determination of the blue metal footprint in forage crops of the DR- 017, Comarca Lagunera, Mexico. Rev. FCA UNCUYO, 2015. 47(1): 93-107- ISSN print 03704661. ISSN (online) 1853-8665, Mendoza, Argentina

**Rios, Torres and Azpilcueta, 2017.** *Op. cit.*

comes from dividing the AFP of the crop to be contrasted by the AFP of the MT apple) greater than 1, to be exact its index was 1.39, which suggests that it produces 39% more apple per m$^3$ of water used in production than the state average of MT apple at the level of the whole state of Chihuahua.

**Table 11. Productivity and physical, economic and social efficiency of water used in the <u>production of MT apple from Chihuahua versus different fruit trees. MT apple = 1.00</u>**

| Product | Place | PFA (kg m'$^3$) | ADP (USD gain hm'$^3$) | EAP (Jobs hm'$^3$) | EFA (L kg') | EEA (m$^3$/USD gain) | ESA (m$^3$/employment) | Author |
|---|---|---|---|---|---|---|---|---|
| **Apple MT** | Chihuahua | 2.28 | $619,570 | 23.5 | 438 | 1.61 | 42,562 | This work |
| **Apple MT** | Chihuahua | 1.00 | 1.00 | | 1.00 | 1.00 | 1.00 | |
| **Pecan nut MT** | Chihuahua | 0.07 | 0.2 | 0.38 | 15.07 | 4.88 | 2.65 | This work |
| **Apple BT** | Cuauhtemoc, Chihuahua | 0.4 | 0.14 | 1.2 | 2.51 | 7.35 | 0.84 | Rios, Torres y Azpilcueta, 2015 |
| **Apple AT** | Cuauhtemoc, Chihuahua | 1.39 | 0.99 | 0.97 | 0.72 | 1.01 | 1.04 | Rios, Torres y Azpilcueta, 2015 |
| **Apple MT** | Cuauhtemoc, Chihuahua | 0.89 | 0.46 | 0.83 | 1.13 | 2.18 | 1.2 | Rios, Torres y Azpilcueta, 2015 |
| **Apple BT** | Canatlan, Dgo. DDR | 0.39 | 0.24 | 1.19 | 2.6 | 4.17 | 0.84 | Rios *et al*, 2015 |
| **Apple AT** | Canatlan, Dgo DDR | 0.82 | 0.59 | 1.64 | 1.21 | 1.71 | 0.61 | Rios *et al*, 2015 |
| **Apple** | Santiago Papasquiaro, Dgo | 0.23 | 0.04 | 1.14 | 4.36 | 25.69 | 0.87 | Navarrete et *al,2015* |
| **Average Pecan Walnut in B y G** | Delicias, Chihuahua | 0.05 | 0.08 | 0.17 | | 0.61 | 5.97 | Rios, Torres y Torres, 2016 |
| **Average Pecan Walnut in B y G** | Southwest Coahuila | 0.03 | 0.02 | 0.72 | 35.89 | 56.73 | 1.38 | Rios y Navarrete, 2017 |
| **Pecan walnut** | Torreon, Coahuila | 0.03 | 0.05 | 0.73 | 35.5 | | 1.37 | Rios y Navarrete, 2017 |
| **Datil** | Comondu, BCS | 0.05 | 0.13 | 0.16 | 22.11 | 7.43 | 6.12 | Zamora y Rivas, 2019 |
| **Olive tree** | Spain | 0.17 | 1.72 | | | | | Montesinos *et al*, 2011 |

| | | | | |
|---|---|---|---|---|
| **Apple** | World average | 0.53 | 1.88 | Mekonnen y Hoekstra, 2011 |
| **State production in general** | California, USA | | 0.4 | Fulton, Cooley y Gleick, 2012 |

**Source:** Own elaboration, based on tables 4 and 9.

All the other crops in table 11 had an index lower than 1.00, i.e. they produce less biomass per m$^3$ of water used in the production than the MT apple from the state of Chihuahua with 2.28 kg m-3. Thus, for example, for the case of the different apples from Chihuahua and Durango, table 11 shows that the lowest of the indices was for the apple produced in the DDR of Santiago Papasquiaro[64] , with 0.23, that is to say, that in Santiago Papasquiaro only 23% of the 2.28 kg produced by the MT apple is produced.28 kg produced by the MT apple of Chihuahua per m$^3$ , and the highest of the indices was for the MT apple of Cuauhtemoc[65] with 89% (the index was 0.89) of the level of biomass produced by the average MT apple of the state, the three walnut trees indicated in table 11 produce between 3 and 5% of biomass per m$^3$ of water than the average MT apple of Chihuahua.

The PFA of date[66] , olive[67] and the world average apple[68] in table 11 produce less kg of biomass per m$^3$ of water than the average MT apple in the state of Chihuahua, as their indices 0.05, 0.17 and 0.53 suggest that they produce only 5%, 17% and 53% of the biomass produced by the MT apple in this study.

Against the average PFA of beef from China, India, USA and the world average determined by Mekonnen and Hoekstra (2012)[69] with 0.073 kg m$^{-3}$ , 0.060 kg m m$^{-3}$ 0.070 kg m-3 and 0.065 kg m-3 respectively, in relation to these livestock products,

---

[64] **Navarrete *et al, 2015*.** *Op. cit.*
[65] **Rios, Torres and Azpilcueta, 2017**. *Op. cit.*
[66] **Zamora and Rivas, 2019.** *Op. cit.*
[67] **Montesinos *et al,* 2011.** *Op. cit.*
[68] **Mekonnen and Hoekstra, 2011.** *Op. cit.*
[69] **Mekonnen, M.M. & Hoekstra, A. G. (2012).** A global assesment of the water footprint of farm animal products. ECOSYSTEM (2012). 15:401-415. DOI:10.1007-s10021-011-9517-8

the PFA of MT apple and MT walnut from Chihuahua (with 2.28 and 0.15 kg m$^{-3}$ respectively), it is observed that both crops enjoy a higher PFA than these livestock products. Similarly, it is observed that of both crops only the MT apple in this study has a higher AFP than the AFP of bovine milk from the Specialized System in Delicias (with 0.220 kg m$^{-3}$ , Rios, Rios and Rios, 2019)[70] and bovine milk from the Specialized system in La Laguna (with 3.344 m$^3$ of water per litre of milk equivalent to 0.309 kg m$^{-3}$ , Rios and Ruiz, 2019)[71] .

The economic productivity of water, or simply EPP, standardized to the unit of measurement of USD profit per cubic hectometre, abbreviated as USD hm-3, as shown in Table 11, was USD 619,570 hm$^3$ = 1.00 in the Chihuahua MT block, noting that only the Olive tree[72] from Spain produces more profit than the reference crop, as its indicator, equal to 1.72, suggests that the same hm$^3$ of water used in the MT apple, produces 72% more profit, but not all the other crops in table 11, with indices lower than 1.00, suggesting that all these crops produce less profit per hm$^3$ than the average MT apple in the state of Chihuahua. On the other hand, its inverse, the FSS, measured in m$^3$ of water per USD of profit produced, abbreviated as m$^3$ /USD, equal to 1.61 m$^3$ /USD=1.00 in the reference crop, the MT apple, when contrasted against the crops indicated in table 11, it is observed that only the already mentioned MT apple from Cuauhtemoc, Chihuahua is similar in its FSS with 1.01, while the other indices, all higher than 1.00, suggest that these crops, all of them, demand more water than the MT apple from Chihuahua to produce one USD of profit, the most extreme case being the walnut tree in southwestern Coahuila[73] , a crop

---

[70] **Rios-Flores, J. L., Rios A., Becky E., Rios A., Hebrian E. 2019**. Physical and economic footprints of milk. The case of bovine milk from Delicias, Chihuahua, Mexico. Editorial Academica Espanola. ISBN 978-620-0-02518-0. Beau Bassin, Mauritius

[71] **Rios-Flores, Jose Luis; Ruiz Torres, Jose. 2019**. Economic productivity of water in agriculture and dairy cattle in Delicias, Chihuahua, Mexico. CiBIyT Journal. Basic Sciences, Engineering and Technology. ISSN 1870-056X. Tlaxcala, Mexico. Pp.46-51.

[72] **Montesinos et al, 2011**. *Op. cit.*

[73] **Rios and Navarrete, 2017**. *Op. cit.*

that demands 56.73 times the amount of water demanded by the MT apple from Chihuahua to produce one US dollar of profit.

The social water productivity (SWP), measured under the units of jobs generated per cubic hectometre of water used in production, abbreviated as jobs $hm_{-3}$, in Table 11, where the unit is the indicator for manzana MT, with 23.5 jobs $hm^{-3}$ , shows that the BT apple from, the BT apple from, the AT apple from and the Santiago Papasquiaro apple had indices higher than unity, i.e., that these crops generate more jobs per volumetric unit of water used in production, while the MT pecan nut crops of the AT and MT apple from Cuauhtemoc, Chihuahua, as well as the three pecan walnut trees indicated had a lower PES than the MT apple from Chihuahua.

The inverse index of PES is that of SWE, measured as m3 of water used in production for each job generated, shows that 42,562 $m^3$ of water were necessary to create one job in the MT manzana MT of Chihuahua (equal to 1.00), and it is observed in table 11 that the crops with a social efficiency of water higher than that of the MT apple tree were the BT apple of Cuauhtemoc (with an index of 0.84), the BT and AT apple of Canatlan (with indices of 0.82 and 0.61) and the Papasquiaro apple (with an index equal to 0.87) as their indices were less than 1.00, while the crops with the lowest social water efficiency (as they use more water to create one job) were the MT pecan nut of this study (a crop that demands 2.66 times the water demanded by the MT apple to create one job), the AT and MT apples of Cuauhtemoc, the three pecan walnuts and the BCS date, it is insisted, their indices indicate that they use more water than the MT apple to create one job.

# VI. CONCLUSIONS AND RECOMMENDATIONS

## 6.1. Conclusions

The objectives of determining the productivity and efficiency of water used in the production of MT pecan nut and contrasting it with the corresponding indicators for MT apple, both crops at the state level of Chihuahua, were fulfilled.

Based on the results of the mathematical models used, the first hypothesis is accepted, since the physical water productivity "PFA" of MT apple (2.28 kg m$^{-3}$ ) was 14.065 times (=15.065 -1, see table 9) higher than the PFA of MT walnut (0.15 kg m$^{-3}$ ) in the state of Chihuahua.

Based on the results of the mathematical models used, the second hypothesis is accepted, since the economic water productivity "EAP" of the MT apple (USD 0.62 profit per m$^3$ ) was 388% (=4.88 - 1, see table 9) higher than the EAP of the MT walnut (USD 0.13 profit per m$^3$ ) in the state of Chihuahua.

Based on the results of the mathematical models used, the third hypothesis is accepted, since the social water productivity "WSP" of MT apple (23 jobs per hm$^3$ of water used in production) was 166% (=2.66 -1, see table 9) higher than the WSP of MT walnut (9 jobs per hm$^3$ of water used in production) in the state of Chihuahua.

## 6.2. Recommendations

The population has been increasing exponentially over the last century, which has led to a concomitant exponential increase in the demand for food, especially due to the effect of the production of fodder to feed livestock dedicated to the production of both milk and meat, although the food of vegetable origin that the population demands directly has also been growing exponentially, which has necessarily brought with it the impact of exerting considerable pressure on the natural resources used in their production, This has necessarily brought with it the impact of exerting a notorious pressure on the natural resources used in their production, water being one of those natural resources under enormous pressure, if not the most, so that, necessarily and inevitably, the optimisation of the efficient and productive use of water used in production, It is **advisable to** have scientific studies that show the degree of efficiency and the degree of productivity with which water is used in agricultural production, in order to obtain numerical indicators by means of which decisions can be made regarding which crops it is more convenient, in terms of sustainability, to allocate water use, since, by having indicators that clearly indicate the physical, economic and social efficiency and productivity of the water used in agricultural production, it will be possible to obtain a more accurate picture of the efficiency and productivity of the water used in agricultural production, Only by having indicators that clearly indicate the physical, economic and social efficiency and productivity of the water used in production, and only by having them, will it be possible, it is insisted, to allocate water to the crops that are both the most environmentally friendly in terms of the ecological environment in that they require little water to produce one kg of physical product, and also the most productive in terms of water use, in that they are those that generate the most profit and employment per cubic metre of water used in their production.

It is recommended that such studies on the water footprint (through productivity and water efficiency indices) of apple cultivation in other producing states such as Chihuahua, Coahuila, Nuevo Leon and Puebla, as these states have different production systems as well as different apple varieties that may have different effects on productivity and water use efficiency, i.e. on the water footprint of the crop.

***Comite Mexicano del Sistema Productivo Nuez A.C. 2018. Y Alderete y Socios, Industrial Consultancy. 2018.*** *Pecan nut strategic study. Actualization 2018. Available  at:.* http://comenuez.com/wp-content/uploads/2018/assets/estudio-estrategico-nuez-pecanera--2018.pdf. *Last accessed: 12 February, 2020.*

**Agua.org.mx** Fondo para la Comunicacion y la Education Ambiental A.C. Overview of water in Mexico. Available at: https://agua.org.mx/cuanta-agua-tiene- mexico/. Last Accessed: January 21, 2020

**Currency converter. 2020.** Available at: https://www.xe.com/es/currencyconverter/convert/?Amount =1&From=USD&To=MXN

**Carrillo, C. J. 2019.** Economic-social productivity of water in drip-irrigated grapevine (*Vitis vinifera)* in Coahuila, Mexico. Professional thesis. Unidad Regional Universitaria de Zonas Aridas, Universidad Autonoma Chapingo. Bermejillo, Durango, Mexico

**Number of inhabitants on the planet.** Available at: mmm https://www.google.com/search?q=cantidad+of+inhabitant+on+t he+planet+in+2018&rlz=1C1AVNC_inMX648MX669& oq=amount+of+inhabitant+on+the+planet+in+2018&aqs=c hrome..69i57j0l3.7834j0j8&sourceid=chrome&ie=UTF-8

**Cifuentes Rodriguez, Reynau, 2017**. Agricultural productivity of water, soil, capital and labour in the cultivation of walnut (*Carya illinoensis*) in San Pedro, Coahuila. Professional thesis. Department of Irrigation. Universidad Autonoma Agraria Antonio Narro, Laguna Unit. Torreon, Coahuila, Mexico.

**SIAP, 2020.** Closing     agncola 2018. Available in: https://nube.siap.gob.mx/cierreagricola/

**Consejo Nacional de Poblacion y Vivienda, 2019.** Available at: https://www.google.com/search?source=hp&ei=c7_aXc34J8 mosgXSxKmwDQ&q=rate+of+growth+demogr%C3%

A1fico+in+mexico&oq=rate+of+growth+demogr%C3%
%A1fico+in+mexico&gs_l=psy-
ab.3...1744.10190..10671...0.0..0.166.166.0j10 1..gws
-wiz0        .me-
qtiFm9a0&ved=0ahUKEwiN_smesYPmAhVJlKwKHVJJiCtY
Q4dUDCAY&uact=5

**Ecosarga**, no date. The properties of apples according to
https://www.frutadelasarga.com/blog/las-propiedades-de- the-
apples-according-to-their-colour
your colour. Available in:

**Evolution of the world population**. The market economy:
virtues and innovations. Demograffa. 2019. Available in:
http://www.juntadeandalucia.es/averroes/centros-
tic/14002996/helvia/aula/archivos/repositorio/250/271/html/e
conomia/2/evolucion.htm. Accessed on: September 10, 2019

**FAO (Food and Agriculture Organisation of the United
Nations), Agriculture and Consumer Protection
Department, 2005.** Water use in agriculture. Available at:
http://www.fao.org/ag/esp/revista/0511sp2.htm

**FIRA, 2020.** Agrocost. Available at:
https://www.fira.gob.mx/Nd/Agrocostos.jsp

**Aquae Foundation, 2019.** Quantity of drinking water, source of
life. Available at: https://www.fundacionaquae.org/wiki-
aquae/datos-datos-del-agua/cantidad-de-agua-potable-source-
of-life/ date accessed: 10 September, 2019.

**Fulton, Julian; Cooley, Heather and Gleick, Peter H. 2012.
California's Water Footprint.** Pacific institute ISBN 1-
893790-46-0 and ISBN 13-978-1-893790-46-9. Oakland,
California. Available in:
https://indicators.ucdavis.edu/water/files/California%20Wate
r%20Footprint%202012%20Pacific%20Institute%20Fulton%
20et%20al.pdf . last accessed: February 9, 2020.

**Gleick (2000).** The World's Water, 2000-2001: The Biennial
Report on Freshwater Resources. Washington, DC. Islan

Press, 2000. 335p.

**HOEKSTRA, A. Y. and HUNG, P.Q. (2005)**. "Globalization of water resources: international virtual water flows in relation to crop trade". Global Environmental Change, 15, pp. 45-56.

**INIFAP-CENID-RASPA. (2006).** *Irrigation Programme.* [Accessed on: 1 September 2019]. Available at: https://cenidraspa.org/serg/serg_v1.php

**The six surprising properties of walnuts.** Available at: https://comefruta.es/las-6-sorprendentes- properties-of-nuts

**Mekonnen, M. M. and Hoekstra, A. Y. 2011.** The green, blue and grey water footprint of crops and derived crop products. *Hydrology and Earth System Sciencies,* 15(5): 1577-1600.

**Mekonnen, M. M. & Hoekstra, A. G. (2012).** A global assesment of the water footprint of farm animal products.

**ECOSYSTEM (2012).** 15:401-415. DOI: 10.1007-s10021-011-9517-8

**Mekonnen, M. M. and Hoekstra, A. Y. 2011.** The green, blue and grey water footprint of crops and derived crop products. Hydrology and Earth System Sciencies, 15(5): 1577-1600.

**Mexico-Population, 2018.** Expansion/ datos macro.com. Available at:
https://datosmacro.expansion.com/demografia/poblacion/mexico

**Montesinos, P.; Camacho, E.; Campos, B.; Rodriguez_Diaz. J. 2011.** Analysis of Virtual Irrigation Water. Aplication to Water Resources Management in a Meditterranean River Basin. Water Resources management. 25(6): 1635-1651.

**Murphy D.E. (2003).** In a first, U.S. puts limits on California's thirst. New York Times, 5 January. 1- 16 p.

**Navarrete-Molina, Cayetano, Ruiz-Esparza, Manuel de Jesus Azpilcueta, Rios-Flores, Jose Luis and Torres-Moreno, Marco Antonio. 2017.** Agricultural water productivity in the cultivation of apple (*Malus domestica Borkh)* produced in the municipalities of Santiago Papasquiaro and Canatlan,

Durango, Mexico. In the book: Perez Soto, Francisco; Figueroa-Hernandez, Esther; Godinez-Montoya,
Lucila; Salazar-Moreno, Raquel. Social Sciences: Economics and Humanities. Handbook T-III. Macroeconomic variables in agricultural production. ISBN 978607-8534-29-6 pp. 93-106. Universidad Autonoma Chapingo, July, 2017. Available at: http://www.ecorfan.org/handbooks/

**Ortiz, Ramos Daniel, 2016**. The production and foreign trade of the walnut (*Carya ilinoinensis*) in Mexico. Professional thesis of Licenciado en Comercio Internacional. Universidad Autonoma Agraria Antonio Narro, Unidad Saltillo. Saltillo, Coahuila Mexico Pag. 33. Available at: http://repositorio.uaaan.mx:8080/xmlui/bitstream/handle/123 456789/8211/T19328%20ORTIZ%20RAMOS,%20DANIEL. pdf?sequence=1

**Pedroza, S. A., Rios-Flores, J. L., Torres, M. M., Cantu B. J. E., Piceno S. C., Yanez Ch. L. G. 2014**. Irrigation water efficiency in the production of forage maize (*Zea mays)* and alfalfa (*Medicago sativa*): social and economic impact. Terra Latinoamericana volume 32 number 3, July-September 2014. pp.231-239. Chapingo, Mexico.

What are        Available at: https://www.espn.com.mx/espn-run/note/_/id/2722475/nutrition-what-is-phytochemicals

**Rios-Flores, J. Luis; Torres M., Miriam; Torres M., M. Antonio. 2015**. Determination of the metal footprint in the production of apples in Canatlan, Durango. In the book: Alternativas sustentables de participacion comunitaria para el cuidado del medio ambiente. Ramon Rivera (Coordinator). ISBN 13:978-84-16399-67-3. 1ª edicion Universidad Autonoma Chapingo, Mexico-Universidad de Antioquia, Colombia. pp. 117-128.

**Rios-Flores, J. Luis, Torres M., Miriam, Castro F., Rafael, Torres M., M.A. Ruiz T. Jose. 2015**. Determination of the blue

Mdrica footprint in forage crops of DR017 Comarca Lagunera, Mexico. Rev. FCA UNCUYO, 2018. 47(1): 101-122, ISSN print 0370-4661. ISSN (online) 1853-8665, pp.93-107. Mendoza, Argentina.

**Rios-Flores, Jose Luis, Torres M. Miriam and Azpilcueta RE, M de J. (2017)**. Water productivity in apple trees produced under different levels of technification in Cuauhtemoc, Chihuahua, Mexico. Revista Asuntos Economicos y Administrativos No. 32, Primer semestre 2017. ISSN 0124-1133. University of Manizales, Colombia. pp. 135-146.

**Rios-Flores, Jose Luis, Torres M. M. and Torres M., M. A. (2016)**. Agricultural water productivity of pecan nut trees in northern Mexico. Cases: Comarca Lagunera and Delicias, Chihuahua. ISBN978-3-639-80166-8. Editorial Academica Espanola. BahnhofstraR>e 28, D-66111, SaarbrUcken, Germany.

**Rios-Flores Jose Luis, Torres Moreno Miriam, Ruiz Torres Jose, Torres Moreno Marco Antonio. 2016.** Efficiency and productivity of water irrigation in wheat (*Triticum vulgare*) from Ensenada and Valle de Mexicali, Baja California, Mexico. Acta Universitaria Journal. Multidisciplinary Scientific Journal. 26(1):20-29 ISSN 0188-6266 Vol. 26 No. 1. January-February 2016. Guanajuato, Mexico.

**Rios-Flores, J. Luis and Navarrete-Molina, C. (2017)**. Huella Mdrica y productividad economica del agua en nogal pecanero (Carya illinoensis*)* al sur oeste de Coahuila, Mexico. Journal: Estudios de Econom^a Aplicada. Volume 35-3, September 2017. ISSN 1133-3197. International Association of Applied Economics (ASEPELT), Spain.

**Rios-Flores, Jose Luis; Ruiz-Torres, Jose; Rios-Arredondo, Becky Elizabeth. 2018.** Economic-social productivity of water in walnut (*Carya illinoensis).* Case study: Walnut production in the municipality of Torreon, Coahuila, Mexico. Editorial Academico Espanola. ISBN 978-620-213967-0. Beau Bassin,

Mauritius.

**Rios-Flores, Jose Luis, Rios Arredondo, Becky Elizabeth, Cantu Brito, Jesus Enrique, Rios Arredondo, Hebrian Efrain, Armendariz Erives, Sigifredo, Chavez Rivero, Jose Antonio, Navarrete Molina, Cayetano & Castro Franco, Rafael (2018)**. Analisis de la eficiencia ffsica, economica y social del agua en esparrago (*Asparagus officinalis L.)* y uva (*Vitis* vinifera) mesa del DR-037 Altar- Pitiquito-Caborca, Sonora, Mexico 2018. *Revista de la Facultad de Ciencias Agrarias. National University of Cuyo*, 50(2). ISSN in print 0370-4661, ISSN (online) 1853-8665. Mendoza, Argentina.

**Rios-Flores, Jose Luis; Rios Arredondo, Becky Elizabeth; Rios Arredondo, Hebrian Efram. 2019**. Physical and economic footprints of milk. The case of bovine milk from Delicias, Chihuahua, Mexico. Editorial Academica Espanola. ISBN 978-620-0-02518-0. Beau Bassin, Mauritius

**Rios-Flores, Jose Luis; Ruiz Torres, Jose. 2019**. Economic productivity of water in agriculture and dairy cattle in Delicias, Chihuahua, Mexico. CiBlyT Journal. Basic Sciences, Engineering and Technology. ISSN 1870-056X. Tlaxcala, Mexico. Pp.46-51.

**Rios-Flores. J. L., Ruis Torres, J., Azpilcueta Ruiz Esparza, M. de J. 2019**. Economic-social productivity of water in avocado cultivation. The case of production in Michoacan, Mexico. Editorial Academico Espanola. ISBN 978-3-639-53183-1. Beau Bassin, Mauritius

**SAGARPA, 2017**. Planeacion Agricola Nacional 2017-20130. Manzana Mexicana. Available at: https://www.gob.mx/agricultura/acciones-y-programas/planeacion-agricola-nacional-2017-2030-126813.

**SIAP, 2017.** Apple: Mexico produced 716,930 tonnes of apples in 2016. Available at: https://www.gob.mx/siap/articulos/manzana-mexico- produced-716-930-tons-in-2016?idiom=es

**SIAP, 2018.** Apple statistics in Mexico. https://blogagricultura.com/estadisticas-manzana-mexico/. Last accessed: 27 January, 2020

**SIAP-SADER (2018). Agricultural closure 2018.** Available at: http://infosiap.siap.gob.mx/aagricola_siap_gb/icultivo/

**Takele, E. and Kallenbach, R. (2001).** Analysis of the Impact of Alfalfa Forage Production under Summer Water-Limiting Circumstances on Productivity, Agricultural and Growers Returns and Plant Stand. Journal of Agronomy and Crop Science, 187 (1): 41-46.

**Zamora, Ramirez, Anselmo & Rivas, L., Jose Antonio. 2019.** Physical, economic and social productivity of water used in the production of datil (*Phoenix dactylifera L.*) irrigated by gravity in Comondu, Baja california Sur. Professional thesis. Universidad Autonoma Chapingo, Unidad Regional Universitaria de Zonas Aridas. Bermejillo, Durango, Mexico.

# Annex

Annex 1: Rural Development Districts (RDD), Rural Development Support Centres (CADER) y municipalities < in apple MT y walnut MT in Chihuahua, Mexico in 2018. The following are taken into account in the harvested areas y its annual physical production

| DDR | CADER | Municipality | Harvested area (ha) | Production (ton) | VBP (nominal MX$) |
|---|---|---|---|---|---|
| Apple MT | | | | | |
| Casas Grandes | Nuevo Casas Grandes | Casas Grandes | 247.38 | 5290 | $47,481,770.40 |
| Wood | Wood | Wood | 105.26 | 2072 | $28,754,600.00 |
| Wood | Gomez Farias | Gomez Farias | 15.29 | 336 | $4,267,200.00 |
| Cuauhtemoc | Anahuac | Cuauhtemoc | 4858.84 | 99991.46 | $1,288,388,087.00 |
| Cuauhtemoc | Cusihuiriachi | Cusihuiriachi | 2064.46 | 53421.96 | $718,183,417.70 |
| Cuauhtemoc | Bachiniva | Bachiniva | 1970.82 | 38718.13 | $536,996,228.90 |
| Papigochi | The Board | Guerre ro | 3035.81 | 61080 | $635,969,540.90 |
| Papigochi | Guerre ro | Guerrero | 4247.62 | 78826 | $828,802,726.30 |
| Papigochi | Matachi | Temosachic | 54.53 | 1125 | $10,349,538.75 |
| | TOTAL | | 16,600.01 | 340,860.55 | $4,099,193,109.95 |
| Pecan nut MT | | | | | |
| Casas Grandes | Nuevo Casas Grandes | Casas Grandes | 1246.25 | 1844.1 | $148,549,981.80 |
| Buenaventura 5 | Buenaventura | Buenaventura | 1188 | 1799 | $146,118,773.80 |
| Buenaventura | Buenaventura | Galeana | 2576 | 3879 | $320,878,793.20 |
| El Carmen | El Carmen | Buenaventura | 2370 | 4218.6 | $343,509,587.50 |
| El Carmen | Villa Ahumada | Ahumada | 2640 | 4672.8 | $373,768,440.40 |
| Vallede Juarez | San Isidro Praxedis G. | Juarez | 938 | 1407 | $104,604,554.70 |
| Vallede Juarez | Guerrero Praxedis G. | Guadeloupe Praxedis G. | 99.75 | 149.63 | $11,205,004.78 |
| Vallede Juarez | Guerrero | Guerrero | 65 | 104 | $7,880,004.08 |
| Lower Rio Conchos | Coyame | Coyame del Sotol | 1000 | 1680 | $139,960,800.00 |
| Lower Rio Conchos | Ojinaga | Ojinaga | 400 | 688 | $56,844,500.16 |
| Lower Rio Conchos | M. Benavides | Manuel Benavides | 28 | 44.24 | $3,402,800.12 |
| Balleza | Balleza | Balleza | 380 | 589 | $44,419,994.55 |
| Delicacies | Saucillo | The Cross | 1661.39 | 2924.05 | $250,152,340.00 |
| Delicacies | Saucillo | Saucillo | 4184.45 | 7280.94 | $619,716,881.00 |
| | TOTAL | | 18,776.84 | 31,280.36 | $2,571,012,456.09 |

Source: Own elaboration, based on "cierre agricola 2018", from SIAP, 2019.

Printed by Books on Demand GmbH, Norderstedt / Germany